Annals of Mathematics Studies

Number 30

ANNALS OF MATHEMATICS STUDIES

Edited by Emil Artin and Marston Morse

1. Algebraic Theory of Numbers, *by* HERMANN WEYL
3. Consistency of the Continuum Hypothesis, *by* KURT GÖDEL
6. The Calculi of Lambda-Conversion, *by* ALONZO CHURCH
7. Finite Dimensional Vector Spaces, *by* PAUL R. HALMOS
10. Topics in Topology, *by* SOLOMON LEFSCHETZ
11. Introduction to Nonlinear Mechanics, *by* N. KRYLOFF *and* N. BOGOLIUBOFF
14. Lectures on Differential Equations, *by* SOLOMON LEFSCHETZ
15. Topological Methods in the Theory of Functions of a Complex Variable, *by* MARSTON MORSE
16. Transcendental Numbers, *by* CARL LUDWIG SIEGEL
17. Problème Général de la Stabilité du Mouvement, *by* M. A. LIAPOUNOFF
19. Fourier Transforms, *by* S. BOCHNER *and* K. CHANDRASEKHARAN
20. Contributions to the Theory of Nonlinear Oscillations, Vol. I, *edited by* S. LEFSCHETZ
21. Functional Operators, Vol. I, *by* JOHN VON NEUMANN
22. Functional Operators, Vol. II, *by* JOHN VON NEUMANN
23. Existence Theorems in Partial Differential Equations, *by* DOROTHY L. BERNSTEIN
24. Contributions to the Theory of Games, Vol. I, *edited by* H. W. KUHN *and* A. W. TUCKER
25. Contributions to Fourier Analysis, *by* A. ZYGMUND, W. TRANSUE, M. MORSE, A. P. CALDERON, *and* S. BOCHNER
26. A Theory of Cross-Spaces, *by* ROBERT SCHATTEN
27. Isoperimetric Inequalities in Mathematical Physics, *by* G. POLYA *and* G. SZEGO
28. Contributions to the Theory of Games, Vol. II, *edited by* H. KUHN *and* A. W. TUCKER
29. Contributions to the Theory of Nonlinear Oscillations, Vol. II, *edited by* S. LEFSCHETZ
30. Contributions to the Theory of Riemann Surfaces, *edited by* L. AHLFORS *et al.*
31. Order-Preserving Maps and Integration Processes, *by* EDWARD J. MCSHANE
32. Curvature and Betti Numbers, *by* K. YANO *and* S. BOCHNER

CONTRIBUTIONS TO THE THEORY OF RIEMANN SURFACES

Centennial Celebration of Riemann's Dissertation

L. V. AHLFORS
S. BERGMAN
L. BERS
S. BOCHNER
E. CALABI
L. FOURÈS
M. HEINS
J. A. JENKINS
S. KAKUTANI
W. KAPLAN
K. KODAIRA
M. MORSE
Z. NEHARI
P. C. ROSENBLOOM
H. L. ROYDEN
L. SARIO
A. C. SCHAEFFER
M. SCHIFFER
M. SHIFFMAN
D. C. SPENCER
S. E. WARSCHAWSKI

Edited by

L. Ahlfors, E. Calabi, M. Morse, L. Sario, D. Spencer

Princeton, New Jersey
Princeton University Press
1953

London: Geoffrey Cumberlege, Oxford University Press
L. C. Card 51-23145

Printed in the United States of America
by Westview Press, Boulder, Colorado

Princeton University Press On Demand Edition, 1985

FOREWORD

On December 26, 1851, Bernard Riemann presented his inaugural dissertation. On December 14, 15, 1951, a Conference on Riemann Surfaces was held in Princeton, New Jersey in commemoration of this event. The Institute for Advanced Study and Princeton University acted as joint sponsors. The present volume of the Annals of Mathematics Studies contains the papers presented at this conference, and the papers are published in the order in which they were presented.

The diversity of mathematical interest and approach apparent in the papers presented here is a small indication of the breadth of influence of Riemann's ideas upon modern mathematics.

Committee on Arrangements
Marston Morse, Chairman
Lars V. Ahlfors
Eugenio Calabi
Leo Sario
D. C. Spencer

CONTENTS

CONTRIBUTIONS TO THE THEORY OF RIEMANN SURFACES

DEVELOPMENT OF THE THEORY OF CONFORMAL MAPPING
AND RIEMANN SURFACES THROUGH A CENTURY

LARS V. AHLFORS

This conference has been called to celebrate the hundredth anniversary of the presentation of Bernhard Riemann's inaugural dissertation "Grundlagen für eine allgemeine Theorie der Functionen einer veränderlichen complexen Grösse".

Very few mathematical papers have exercised an influence on the later development of mathematics which is comparable to the stimulus received from Riemann's dissertation. It contains the germ to a major part of the modern theory of analytic functions, it initiated the systematic study of topology, it revolutionized algebraic geometry, and it paved the way for Riemann's own approach to differential geometry.

To deal with all these aspects in the short time at our disposal would be impossible. It is therefore necessary to concentrate on the central idea in Riemann's thesis which is that of combining geometric thought with complex analysis. In this introductory lecture my task will be to trace the main lines along which geometric function theory has expanded from Riemann's time to ours.

Geometric function theory

Riemann's paper marks the birth of geometric function theory. At the time of its appearance Cauchy had already laid the foundation of analytic function theory in the modern sense, but its use was not widespread. It is clear that complex integration had introduced a certain amount of geometric content in analysis, but it would be wrong to say that Riemann's ideas were in any way anticipated. Riemann was the first to recognize the fundamental connection between conformal mapping and complex function theory: to Gauss, conformal mapping had definitely been a problem in differential geometry.

The most astonishing feature in Riemann's paper is the breathtaking generality with which he attacks the problem of conformal mapping. He has no thought of illustrating his methods by simple examples which to lesser mathematicians would have seemed such an excellent preparation and undoubtedly would have helped his paper to much earlier recognition. On the contrary, Riemann's writings are full of almost cryptic messages to the future. For instance, Riemann's mapping theorem is ultimately formulated in terms which would defy any attempt of proof, even with modern methods.

Riemann surfaces

Among the creative ideas in Riemann's thesis none is so simple and at the same time so profound as his introduction of multiply covered regions, or Riemann surfaces. The reader is led to believe that this is a commonplace convention, but there is no record of anyone having used a similar device before. As used by Riemann it is a skillful fusion of two distinct and equally important ideas: 1) A purely topological notion of covering surface, necessary to clarify the concept of mapping in the case of multiple correspondence; 2) An abstract conception of the space of the variable, with a local structure defined by a uniformizing parameter. The latter aspect comes to the foreground in the treatment of branchpoints.

From a modern point of view the introduction of Riemann surfaces foreshadows the use of arbitrary topological spaces, spaces with a structure, and covering spaces.

Existence and uniqueness theorems

There is a characteristic feature of Riemann's thesis which should not be underestimated. The whole paper is built around existence and uniqueness theorems. To us, this seems the most natural thing in the world, for this is what we expect from a paper which introduces a new theory. But we must realize that Riemann is, to say the least, one of the earliest and strongest proponents of this point of view. Again and again, explicitly and between the lines, he emphasizes that a function can be defined by its singularities. This approach calls for existence and uniqueness theorems, in contrast to the classical conception of a function as a closed analytic expression. There is no doubt that Riemann's point of view has had a decisive influence on modern mathematics.

Potential theory

Next to the geometric interpretation, the leading mathematical idea in Riemann's paper is the importance attached to Laplace's equation. He virtually puts equality signs between two-dimensional potential theory and complex function theory. Riemann's aim was to make complex function theory a powerful tool in real analysis, especially in the theory of partial differential equations and thereby in mathematical physics. It must be remembered that Riemann was in no sense confined to a mathematical hothouse atmosphere; his broad mind was prone to accept all the inspiration he could gather from his unorthodox, but suggestive, conception of contemporary physics.

Dirichlet's principle

Riemann's proof of the fundamental existence theorems was based on an uncritical use of the Dirichlet principle. It is perhaps wrong to call Riemann uncritical, for he made definite attempts to exclude a degererating extremal function. In any case, if he missed the correct proof, he made up for it by giving a very general formulation. In modern language, his approach, which is clearer in the paper on Abelian integrals, is the following:

Given a closed differential α with given periods, singularities and boundary values, he assumes the existence of a closed differential β such that $\alpha + \beta^*$ (β^* denotes the conjugate differential) has a finite Dirichlet norm. Then he determines an exact differential ω_1, with zero boundary values, whose norm distance from $\alpha + \beta^*$ is a minimum. But this is equivalent to an orthogonal decomposition

$$\alpha + \beta^* = \omega_1 + \omega_2^* \qquad (\omega_1 \text{ exact, } \omega_2 \text{ closed})$$

from which it follows that

$$\alpha - \omega_1 = \omega_2^* - \beta^*$$

is simultaneously closed and co-closed, that is to say harmonic. Hence the existence theorem: there exists a harmonic differential with given boundary values, periods, and singularities.

The easiest way to make the reasoning exact is to complete the differentials to a Hilbert space. Closed and exact differentials can be defined by orthogonality, and in the final step one needs a lemma of Herman Weyl for which there is now a very short proof. Riemann, who did not have these tools, was nevertheless able to choose this beautiful approach which unifies the problems of boundary values, periods, and singularities.

Schwarz-Neumann

Riemann's failure to provide a rigorous proof for the Dirichlet principle was beneficial in causing a flurry of attempts to prove the main existence theorems by other methods. The first to be successful was H. A. Schwarz who devised the alternating method. Minor improvements were contributed by C. Neumann who is most notable as a popularizer of Riemann's ideas. The alternating method proved to be sufficient to dispose of the existence problems on closed surfaces.

The method of Schwarz is a linear method, and in principle it amounts to solving a linear integral equation by iteration. The difficulties in adapting the method to the existence theorems are of a practical nature, but they are considerable. The advantages are that the method is successful, and constructive, but in simplicity and elegance it does not compare, even remotely, with Riemann's method.

Poincaré-Klein

In the next generation the leaders were Poincaré and F. Klein. Their important innovation is the introduction of the problem of uniformization of algebraic and general analytic curves. The study of Riemann surfaces in this light leads to automorphic functions and the use of non-euclidean geometry. The method is extremely beautiful; as H. Weyl puts it, the nature of Riemann surfaces is reflected in the non-euclidean crystal. It has also the advantage of leading to very explicit representations by way of Poincaré's theta-series and their generalizations.

The disadvantage is that the existence proofs are quite difficult. For compact surfaces Poincaré produces a correct proof based on a continuity method which is simple in principle, but technically very involved. It is interesting to note that Poincaré concentrates his efforts on proving the general uniformization theorem. In spite of its generality this theorem would not even include the Riemann mapping theorem. Poincaré has been extremely influential in developing methods which ultimately led to proofs of the mapping and uniformization theorems, but he did not himself produce a complete proof until 1908, having been preceded by Osgood who proved the Riemann mapping theorem in 1900, and Koebe who proved the uniformization theorem in 1907.

Osgoods' proof is very remarkable, because it is so clear and concise and does not leave any room for doubt. It is based on an idea of Poincaré, but Osgood deserves full credit for making the idea work. The proof uses the modular function, and is thus not elementary.

Koebe

The crowning glory was achieved by Koebe when he proved the general uniformization theorem. This is the theorem which asserts that every simply connected Riemann surface is conformally equivalent with the sphere, the disk, or the plane. It immediately takes care of the uniformization of the most general surface, for it is sufficient to map the universal covering surface. As a tool he uses his famous "Verzerrungssatz".

The stage was now set for deeper investigations of the problem of conformal mapping. The standard theorems which concern the canonical mappings of multiply connected regions are from this time. Koebe was an undisputed leader, and Leipzig a center for conformal mapping.

Looking back one cannot help being impressed by Koebe's life work. His methods were completely different from those of his predecessors, and when the initial difficulties were conquered he did not hesitate to attack new problems of ever increasing complexity.

Idee der Riemannschen Fläche

In the classical literature no clear definition of a Riemann surface is ever given. Primarily, the classical authors thought in terms of multiply covered regions with branch-points, but applications to surfaces in space are not uncommon. It is of course true that F. Klein had a general conception of a Riemann surface which is quite close to modern ideas, but his conception is still partially based on geometric evidence.

H. Weyl's book "Die Idee der Riemannschen Fläche", first published in 1913, was the real eye-opener. Pursuing the ideas of Klein it brings, for the first time, a rigorous and general definition of a Riemann surface, and it marks the death of the glue-and-scissors period. The pioneer qualities of this book should not be forgotten. It is a forerunner which has served as a model for the axiomatization of many mathematical topics. For his definition of the abstract Riemann surface Weyl uses the power series approach. The equivalent definition of Radó is perhaps a little smoother, and Radó added the important recognition that the separability is a consequence of the conformal structure.

Weyl was able to base the existence theorems on the Dirichlet principle which had been salvaged by Hilbert. The book is a reminder of how Riemann's original idea still provides the easiest access to the existence theorems. It has exerted a strong implicit influence by its change of emphasis which has led to a strengthening of the ties between the theory of Riemann surfaces and differential geometry.

Topological aspects

The abstract approach to Riemann surfaces, with all its advantages, tends to neglect the covering surface aspect. With the advances made in topology the notions of fundamental group and universal covering space had become thoroughly familiar, and accordingly the case of smooth covering surfaces was well covered in Weyl's book. Stoilow filled the gap with a study of covering surfaces with branch-points. Whyburn completed the work of Stoilow and based it more firmly on pure topology.

Higher dimensions

Finally, the important question of generalizing Riemann's work to several dimensions has made enormous strides in the last decades. The greenest laurels belong to Hodge for his pioneer research on harmonic integrals on Riemannian manifolds. Through his initiative, and the parallel work of de Rham in topology, it was discovered that the problems of Riemann have a significant counterpart on more-dimensional closed manifolds with a Riemannian metric. The existence theorems presented initial difficulties, but it was finally found that Hilbert's integral equation method as well as Riemann's own method of orthogonal projection can both be made successful (Weyl, de Rham, Kodaira). For the purpose of pursuing the function-theoretic analogy Kählerian manifolds have the most desirable properties, and for such manifolds the problem of singularities has been successfully attacked (A. Weil, Kodaira). Recent advances referring to boundary values will be discussed during this conference.

Meromorphic functions

In discussing the generalizations to several variables I have rushed ahead of the chronological order. I will devote the rest of the lecture to progress in the one variable theory after Koebe's active period. In giving so much space to the modern theory of conformal mapping I may be guilty of overemphasizing the questions which I personally happen to know most about, but there are also objective reasons for being partial to the topics which lie nearest to the central theme in Riemann's thesis.

It is time to recall the important advances made in the theory of entire and meromorphic functions about 1925 and shortly thereafter. Borel, Hadamard, and many others had investigated the properties of entire functions by powerful methods which made use of the canonical representations and power series developments. Their results had an air of being quite definitive, but in an almost sensational paper R. Nevanlinna was able to show that surprisingly elementary potential theoretic methods make it possible to push the study of meromorphic functions very much further. Nevanlinna's theory is quite a show-piece of modern mathematics on a classical basis, and it marked a victory of Riemann's method over the methods which go back to Weierstrass. Nevanlinna's theory of meromorphic functions, and perhaps even more the joint work of the two brothers Nevanlinna, represent a revival of geometric function theory in a direction quite different from the one pursued by Koebe. Among the important tools introduced at this time it is sufficient to mention the introduction of harmonic measure, used independently by Carleman, Ostrowski and the Nevanlinnas.

The type problem

Nevanlinna's main results are generalizations of Picard's theorem. Inasmuch as a meromorphic function maps the plane onto a covering surface they have the character of distortion theorems. This point of view leads almost immediately to the problem of type. The problem is to determine whether a simply connected open Riemann surface, ordinarily given as a covering surface of the sphere, can be mapped conformally onto the whole plane or onto a disk (parabolic or hyperbolic case).

Picard's theorem is the only classical theorem of this sort: if the surface fails to cover three points it must be hyperbolic. In its first phase the problem of type centered about generalizations of this theorem. Generally speaking, if the surface is strongly ramified, it tends to be hyperbolic. It is difficult, however, to measure the ramification. In the special case where all branch-points project into a finite number of points the surface can be described in combinatorial terms with the help of a graph introduced by Speiser. There is a rich literature on this special subject which has produced remarkable results, although the gap between sufficient and necessary conditions is still wide, and likely to remain so.

General methods

There is little point in giving a list of individual results related to the type problem. It is more interesting to discuss the methods by which these results were obtained, especially since they are by no means restricted to this particular problem.

On a Riemann surface it is important to consider all metrics whose element of length is of the form $\rho|dz|$, where z is the local uniformizer. The totality of such metrics is a conformal invariant. It is a classical result that only hyperbolic surfaces can carry a metric with constant negative curvature, and it is not difficult to see that the existence of a metric whose curvature is negative and bounded away from zero is sufficient to imply that the surface is hyperbolic. More generally, the same can be expected if the curvature is negative in the mean, but this property is difficult to formulate in the absence of a natural method to prescribe the weights. What one can do is to consider properties in the large which are roughly equivalent to negative curvature. Such properties are expressed through relations between length and area, for instance in the form of isoperimetric inequalities. In such terms it is possible to formulate necessary and sufficient conditions for the type which have proved very useful.

Method of Grötzsch

Comparisons between length and area in conformal mapping, and the obvious connection derived from the Schwarz inequality, had been used before, notably by Hurwitz and Courant. The first to make systematic use of this relation was H. Grötzsch, a pupil of Koebe. The speaker hit upon the same method independently of Grötzsch and may, unwittingly, have detracted some of the credit that is his due. Actually, Grötzsch had a more sophisticated point of view, but one which did not immediately pay off in the form of simple results.

Extremal length

The strip method of Grötzsch has finally given way to the much more flexible and concise considerations of Beurling. Although the method remains essentially the same, the original idea is given a new twist by putting the emphasis on numerical conformal invariants which obey very simple rules of composition and majoration. This makes the theory so easy to apply that many applications become practically trivial. Above all, it leads to a systematic search for extremal metrics which in many cases are associated with important canonical mappings. The details of this theory of extremal length were developed jointly by Beurling and Ahlfors.

Grunsky

In Germany the tradition of geometric function theory was also carried on by H. Grunsky and O. Teichmüller. The former was a student of

E. Schmidt, and his thesis is a truly remarkable piece of work. He shows, rather surprisingly, that some important extremal problems in the theory of conformal mapping can be successfully attacked by the time-honored device of contour integration. I cannot resist quoting the most beautiful of these results. A region of finite connectivity can be mapped, with a proper normalization, on a slit-region bounded by horizontal or vertical slits. If these mapping functions are $p(z)$ and $q(z)$, Grunsky shows that $p - q$ and $p + q$ have important extremal properties. With a given normalization $p - q$ maps the region onto a Riemann surface of smallest area, while $p + q$ yields a schlicht mapping onto a region whose complement has maximal area. What is more, the contours are convex curves, the same for both mappings (up to a translation and reflection). Part of this result must be accredited to Schiffer, for Grunsky failed to notice that $p + q$ is schlicht. In any case, the idea of applying contour integration to functions of the form $p \pm q$ is due to Grunsky. The same idea has been used later by Schiffer, Spencer and the speaker in the case of general Riemann surfaces with analytic contours.

Teichmüller

In the premature death of Teichmüller geometric function theory, like other branches of mathematics, suffered a grievous loss. He spotted the importance of Grötzsch's technique, and made numerous applications of it which it would take me too long to list. Even more important, he made systematic use of extremal quasi-conformal mappings, a concept that Grötzsch had introduced in a very simple special case. Quasi-conformal mappings are not only a valuable tool in questions connected with the type problem, but through the fundamental although difficult work of Teichmüller it has become clear that they are instrumental in the study of modules of closed surfaces. Incidentally, this study led to a better understanding of the role of quadratic differentials which in somewhat mysterious fashion seem to enter in all extremal problems connected with conformal mapping.

The Finnish school

We now turn our attention to the progress made by the Finnish school under the leadership of R. Nevanlinna and P. Myrberg. The original problem of type had been formulated for simply connected surfaces. While this was natural as long as the problem was coupled with the theory of meromorphic functions, it soon turned out to be an artificial restriction in the general theory of conformal mapping. A very obvious way would have been to consider, for arbitrary surfaces, the type of their universal covering surface, but this does not lead anywhere, for except in the very simplest topological cases the universal covering surface is always hyperbolic.

In the study of general Riemann surfaces it is a good policy to pay careful attention to what happens in the case of plane regions, which are of course considered as special Riemann surfaces. With some care it is usually not too difficult to generalize a theory from plane regions to Riemann surfaces, provided that the theory can be expressed in conformally invariant terms. The notion of potential, and the derived notion of capacity, in the precise definition of Wiener, are not strict conformal invariants, but they have related properties. Myrberg observed that a plane region has a Green's function if and only if its complement is of capacity zero, a property which had not been explicitly stated before. From there it was only a short step to introduce the fundamental dichotomy by which an arbitrary Riemann surface is said to be hyperbolic if it has a Green's function, parabolic if it does not.

Simultaneously, Nevanlinna had looked at the same question from another angle. He found that a closed set of zero capacity has always vanishing harmonic measure, and he was led to consider surfaces whose "ideal boundary" has the harmonic measure zero. It was easy to see that this was identical with Myrberg's classification, surfaces with a null-boundary corresponding to the class of parabolic surfaces.

Both authors proceeded to study the theory of Abelian differentials on parabolic surfaces. Nevanlinna produced a very complete theory of Abelian differentials of the first kind, and Myrberg studied in even greater detail the case of hyperelliptic surfaces of infinite genus. These studies had been preceded by work of Hornich who had investigated some examples.

Classification theory

Parabolic surfaces are degenerate in that they share some of the properties of closed surfaces. One notes, however, that there are certain properties of closed surfaces which carry over to some, but not all, parabolic surfaces. As a continuation of the work begun by Myrberg and Nevanlinna there has lately been much discussion of a more general classification theory of Riemann surfaces. This theory studies different kinds of degeneracy. For instance, one may ask whether a surface can carry bounded analytic functions other than constants, whether it can carry bounded harmonic functions, and so on. It is of interest to relate these properties to each other. It has been demonstrated, for instance, that every surface with a non-trivial bounded harmonic function is hyperbolic, but it is not known whether all hyperbolic surfaces have bounded harmonic functions. Many problems of this nature are still unsolved and seem to present quite a challenge.

Variational methods

Very important progress has also been made in the use of variational methods. I have frequently mentioned extremal problems in conformal mapping, and I believe their importance cannot be overestimated. It is evident that extremal mappings must be the cornerstone in any theory which tries to classify conformal mappings according to invariant properties.

Dirichlet's principle, which is the classical variational problem in function theory, has already been discussed and will not be mentioned further. An interesting attempt to utilize the method of calculus of variations was made by Hadamard who determined the variation of the Green's function for regions with a very regular boundary. The next initiative was taken by Löwner whose most spectacular result was the proof of the inequality $|a_3| \leqq 3$ in the theory of schlicht functions. This was the first decisive advance beyond Koebe's distortion theorem, and it is a result of approximately the same depth as the most recent achievements.

As a systematic tool the method of variation was first introduced by Schiffer. It has also been used by Spencer and Schaeffer in their work on schlicht functions. Important results have been obtained, but what counts more is the creation of a new tool which can be applied with comparative ease to problems which definitely do not lie near the surface.

To give an idea of the method I will describe it in quite general terms, without regard to the fact that this is not precisely the form in which it has been used. I wish only to make the principle clear, without entering in details.

The main application is to extremal problems for schlicht conformal mappings. If one wants to solve such a problem the main difficulty one has to cope with is the effective construction of schlicht variations. Suppose that we are dealing with schlicht mappings of a surface Z into a surface W. If f is such a mapping, and δf a variation, how can we make sure that $f + \delta f$ is again schlicht?

The methods used by Schiffer and Spencer-Schaeffer can be interpreted as follows: On W we introduce a new metric of the form

$$ds^2 = |dw + \epsilon h d\bar{w}|^2,$$

where h behaves properly under changes of the local variable. This metric is not conformal with the metric on W, and thus determines a new Riemann surface W^* with the same points. Through f a corresponding metric will be defined on Z, which is such that f defines a conformal mapping of the new surface Z^* into W^*. Now we define $f + \delta f$ through the series of mappings

$$Z \overset{\alpha}{\rightarrow} Z^* \overset{f}{\rightarrow} W^* \overset{\beta}{\rightarrow} W,$$

where α and β are conformal mappings. This is possible only if Z, Z^* and W, W^* are conformally equivalent. Naturally, the conditions for conformal equivalence lead to the vanishing of certain linear functionals $L_i(h)$[1]. Any side-conditions can be satisfied in the same manner, and if

1) More precisely, linear functionals of h and $\bar{h}$.

$F(f)$ is the functional of f that we try to extremalize, the variation δF is likewise a linear functional of h, and the condition for an extremum can be expressed in the form

$$\delta F(h) = \Sigma \lambda_i L_i(h).$$

This is merely a rough sketch, and it is by no means sure that the program could be carried out in anything like this generality. In any case it should be possible to determine the functionals $L_i(h)$, which express the conformal equivalence, explicitly, and something in this direction has already been done by Garabedian and by Schiffer and Spencer. A serious difficulty is connected with the finding of the mappings which satisfy the variational condition, and finally one is faced with the problem of eliminating the solutions which give only local extremes.

Future progress

To mark the end of this lecture, it is perhaps not useless to point to the directions in which future progress can be expected. Geometric function theory of one variable is already a highly developed branch of mathematics, and it is not one in which an easily formulated classical problem awaits its solution. On the contrary it is a field in which the formulation of essential problems is almost as important as their solution; it is a subject in which methods and principles are all-important, while an isolated result, however pretty and however difficult to prove, carries little weight.

Nothing could be more false than to say that classical function theory has solved its problems and has therefore outlived itself. Even without the introduction of completely new ideas the classical problem of modules, vague as it is, and - to mention a more recent example - the investigation of the true role played by Teichmüller's extremal quasi-conformal mappings, are questions which can keep generations busy. The interaction of topology and function theory is likewise a field which has only been scratched on the surface. Above all, it has happened before that the whole outlook on function theory has changed abruptly, and it will happen again. The spirit of Reimann will move future generations as it has moved us.

Remark

This survey has necessarily followed a narrow central line. Since many names and important contributions have been omitted, it must be underscored that such omissions are in no way indicative of relative merit in the eyes of the writer.

VARIATIONAL METHODS IN THE THEORY OF
RIEMANN SURFACES

Menahem Schiffer

1. Introduction

The abelian integrals and their periods are functionals of the closed Riemann surface $\mathfrak{R}$ on which they are defined. Already Legendre established a differential equation for the periods of elliptic integrals considered as functions of the one modulus of the Riemann surface. Fuchs showed that if the surface $\mathfrak{R}$ is realized as a multiple covering of the complex plane, the abelian integrals and their periods may be differentiated with respect to each branch point of the surface and that the derivatives can be expressed in terms of the abelian integrals themselves. He was led to an interesting theory of differential equations for an algebraic surface which becomes, however, rather involved with increasing genus of the surface. By means of the theta-series the relations between the various quantities on a surface can be exhibited explicitly but lead again to a rather formidable formalism. Ritter and Koebe studied directly the continuous variation of canonical domains under variations of the moduli of the surface.

Since the beginning of this century problems of conformal mapping and boundary value problems in potential theory led to the study of functions which are analytic on Riemann surfaces with boundaries. A function on $\mathfrak{R}$ is no more uniquely determined by its singularities alone but depends now intrinsically on its boundary values. However, the study of the analytic functions on a surface $\mathfrak{R}$ with boundary may be readily reduced to the investigation of the two fundamental harmonic functions in $\mathfrak{R}$, the Green's function and the Neumann's function.

The Green's function $G(p,q)$ is harmonic in p, vanishes on the boundary of $\mathfrak{R}$ and has a logarithmic singularity if $p \to q$. The Neumann's function $N(p,q,q_1)$ is harmonic in p, has a vanishing normal derivative on the boundary of $\mathfrak{R}$ and has two logarithmic poles of opposite sign at the points q and q_1. We may consider G and N as functionals of $\mathfrak{R}$ and study them as such. This theory was developed by Hadamard and Volterra in the case of planar domains. In the case of plane simply connected domains both fundamental functions may be expressed in terms of the mapping function of the domain upon the unit circle, and a theory was developed by Löwner studying its variations under some special kinds of variation of the domain.

2. The closing of surfaces with boundary

The theory of open Riemann surfaces, though appearing at first sight more general than that of closed ones, may be reduced to the latter. We associate with a surface $\mathfrak{R}$ its double $\widetilde{\mathfrak{R}}$ such that to each point $p \in \mathfrak{R}$ corresponds a double point $\widetilde{p} \in \widetilde{\mathfrak{R}}$; if z is a permissible uniformizer at $\widetilde{p}$, then $\bar{z}$ is admitted as uniformizer at $\widetilde{p}$. If we are near a boundary continuum of $\mathfrak{R}$, its neighborhood shall be referred to a neighborhood $\mathfrak{R}$ of the real axis in the half-plane, $\operatorname{Im}\{z\} < 0$ while the corresponding neighborhood in the double $\widetilde{\mathfrak{R}}$ shall correspond to the image of $\mathfrak{R}$ with respect to reflection along the real axis. In this way, an analytic continuation from $\mathfrak{R}$ into $\widetilde{\mathfrak{R}}$ is defined and $\mathfrak{R} + \widetilde{\mathfrak{R}}$ forms a closed Riemann surface

Clearly, the Green's function and the Neumann's function of $\mathfrak{R}$ may be continued analytically across the boundary into $\mathfrak{F}$ as odd and even harmonic functions respectively; that is, we may define

$$G(\widetilde{p},q) = -G(p,q) \quad , \quad N(\widetilde{p},q,q_1) = N(p,q,q_1). \tag{2.1}$$

Let $\Omega_{q\,q_o}(p)$ be the integral of the third kind on $\mathfrak{F}$ which has a single-valued real part on $\mathfrak{F}$ and the logarithmic poles at q and q_o. It is easily seen that

$$G(p,q) = 1/2\left[\Omega_{q\widetilde{q}}(p) - \Omega_{q\widetilde{q}}(\widetilde{p})\right] = \operatorname{Re}\left\{\Omega_{q\widetilde{q}}(p)\right\} \quad . \tag{2.2}$$

and

$$N(p,q,q_1) = 1/2\left[\Omega_{q\widetilde{q}}(p) + \Omega_{q\widetilde{q}}(\widetilde{p}) - \Omega_{q_1\widetilde{q}_1}(p) - \Omega_{q_1\widetilde{q}_1}(\widetilde{p})\right] \quad . \tag{2.3}$$

Thus, the fundamental functions of an open surface are easily expressed in terms of the abelian integrals of the closed surface obtained by adding to it its double. The entire variational theory of these functions becomes more transparent if treated from this point of view and the variational formulas obtained in the case of closed surfaces are at the same time simpler and more general than those valid for the open case.

It should also be observed that the classical results on the existence of abelian integrals on closed surfaces provide new existence proofs for the Green's and Neumann's functions of Riemann surfaces with boundary and guarantee the possibility to solve boundary value problems of the first and second kind for such open surfaces.

3. The fundamental variational formula

We describe now the simplest variation of a closed Riemann surface which we will consider. Let t be a point on $\mathfrak{R}$ and let $z(t)$ be a local uniformizer at t. We choose a circle of radius ρ in the z-plane which lies entirely in the parameter neighborhood and map the whole neighborhood conformally by

$$z^* = z + \frac{e^{2i\varphi}\rho^2}{z}, \quad 0 \leq \varphi < 2\pi. \tag{3.1}$$

This mapping is univalent for $|z| > \rho$ and identifies analytically points on the periphery of the circle $|z| = \rho$ which lie on the same normal to the diameter of the circle in the direction $e^{i\varphi}$. We define as $\mathfrak{R}^*$ the Riemann surface obtained from $\mathfrak{R}$ by identifying the points on the uniformizer circle $|z| = \rho$ in the way described.

Let $W(p,p_o;\ q,q_o)$ be the integral of the third kind on $\mathfrak{R}$ which is symmetric in its argument pairs and has vanishing periods for the odd cycles of a canonical cut system on $\mathfrak{R}$. Let W^* denote the corresponding integral with respect to $\mathfrak{R}^*$. We have then the asymptotic formula:

$$\begin{aligned}\delta W &= W^*(p,p_o;\ q,q_o) - W(p,p_o;\ q,q_o) \\ &= e^{2i\varphi}\rho^2 W'(t,t_o;\ p,p_o)W'(t,t_o;\ q,q_o) + o(\rho^2)\end{aligned} \tag{3.2}$$

where t_o is an arbitrary point on $\mathfrak{R}$ and the dash denotes the derivative of W with respect to its first argument.

It is well-known that the period of $W(p,p_o;\ q,q_o)$ is $U_\mu(q) - U_\mu(q_o)$ if the point p describes the cycle $k_{2\mu}$; here $U_\mu(q)$ is the μ-th integral of the first kind of $\mathfrak{R}$ in the transcendent normalization distinguishing the cycles $k_{2\mu-1}$. Thus, we derive from (3.2) the variational formula:

$$\delta\,[U_\mu(q) - U_\mu(q_o)] = e^{2i\varphi}\rho^2 U'_\mu(t)W'(t,t_o;\ q,q_o) + o(\rho^2). \tag{3.3}$$

If we denote by $P_{\mu\nu}$ the period of $U_\mu(q)$ over a cycle $k_{2\nu}$, we derive from (3.3):

$$\delta P_{\mu\nu} = e^{2i\varphi}\rho^2 U'_\mu(t)U'_\nu(t) + o(\rho^2) \tag{3.4}$$

Since we can construct all other functionals on the surface $\mathfrak{R}$ in terms of the integrals of the third kind, its derivatives and its periods, the formula (3.2) contains the whole variational theory of $\mathfrak{R}$ if we still add that the asymptotic terms $o(\rho^2)$ can be uniformly estimated in each subdomain of $\mathfrak{R}$ which lies outside a neighborhood of t and this remainder term may be differentiated with respect to p and q without a change in its order of magnitude.

The structure of the variational formula (3.2) suggests the introduction of the bilinear differential

$$\frac{\partial^2 W(p,p_0;\ q,q_0)}{\partial p\,\partial q} = L(p,q) \tag{3.5}$$

which leads to the elegant variational result

$$\delta L(p,q) = e^{2i\varphi}\rho^2\ L(t,p)\ L(t,q) + o(\rho^2). \tag{3.6}$$

The bilinear differential $L(p,q)$ with this particularly simple variational behavior plays an important role in various problems of conformal mapping and in the theory of Riemann surfaces. It is closely related to Bergman's kernel function and leads to important criteria for the possibility of imbedding one Riemann surface into another.

If $\mathfrak{R}$ is an open Riemann surface and if we want to study the variation of its Green's and Neumann's functions, we may make use of the preceding results. We have to take care, however, that the surface $\mathfrak{F} = \mathfrak{R} + \tilde{\mathfrak{R}}$ remains a symmetric surface $\mathfrak{F}^* = \mathfrak{R}^* + \tilde{\mathfrak{R}}^*$ after the variation; this is attained by making the identification on $\mathfrak{R}$ of the form (3.1) and making the corresponding identification at the double $\tilde{t}$ in $\tilde{\mathfrak{R}}$. An easy calculation leads to the result:

$$\delta G(p,q) = \mathrm{Re}\left\{4e^{2i\varphi}\rho^2\ \frac{\partial G(t,p)}{\partial t}\ \frac{\partial G(t,q)}{\partial t}\right\} + o(\rho^2), \tag{3.7}$$

$$\begin{aligned}&\delta\,[N(p,q,q_1) - N(p_1,q,q_1)]\\ &= \mathrm{Re}\left\{4e^{2i\varphi}\rho^2\ \frac{\partial N(t,p,p_1)}{\partial t}\cdot\frac{\partial N(t,q,q_1)}{\partial t}\right\} + o(\rho^2)\end{aligned} \tag{3.8}$$

4. Identities between functionals

The variational formulas of the last section contain in principle all identities and differential equations involving the various functionals of a given surface $\mathfrak{R}$. In fact, these formulas may be considered as a parametric representation of all differential relations, the parameters being φ and the point t of the identification. The analytic character of the dependence of the various functionals on given sets of moduli can be studied directly from these equations without appeal to the heavy tools of the theory of the theta-series. The asymmetry of Fuchs' treatment of the differential relations by distinguishing particular realizations and their branch points is also avoided.

Let us illustrate the use of the variational formulas by the following example. Let the abelian integral of the first kind $U(p) = U_1(p)$

serve as local uniformizer on $\mathfrak{R}$; then

$$F(U,V;\mu_j) = \frac{L(p,q)}{U'(p)U'(q)}, \quad V = U(q) \tag{4.1}$$

will be a function on the surface which depends on U,V and the moduli μ_j of $\mathfrak{R}$. It can be shown that if $\mathfrak{R}$ is of genus G a set of $3G-6$ periods $P_{\mu\nu}$ will serve as proper set of moduli and we want to study the character of the function $F(U,V;\, P_{\mu\nu})$. For this purpose, we vary the surface $\mathfrak{R}$ by a variation of the type (3.1); since we know the variations of U, V, L, $U'(p)$, $U'(q)$ and the $P_{\mu\nu}$, and since F is a universal function valid for all Riemann surfaces, we will obtain a differential equation for F. It appears that $F(U,V;\, P_{\mu\nu})$ is a function of the combination $U - V$, and denoting $U(t) = T$, $F(U,V;P_{\mu\nu}) = f(U - V;P_{\mu\nu}) = f(U - V)$ we find:

$$f(T-U)\, f(T-V) - f(U-V)\, [f(T-U) + f(T-V] \tag{4.2}$$

$$= f'(U-V) \int_V^U f(Z-T)\, dZ + \Sigma \frac{\partial f}{\partial P_{\mu\nu}} U'_\mu(T)\, U'_\nu(T).$$

When we specialize to a surface of genus one, this simple relation leads to the Schwarz' differential equation for the elliptic functions in dependence on the modulus and contains in this case also the ordinary differential equation of the Weierstrass' function $\wp(u)$. In fact, in this case the function $F(U)$ turns out to be essentially $\wp(u)$ and the simple structure of variational formula (3.6) may explain why just this particular elliptic function serves best as the starting point for a general theory of elliptic function.

5. Variations changing topological type

All variations changing the topology of a surface may be reduced to the following two deformations:

(a) We choose a point $t \in \mathfrak{R}$ and a fixed uniformizer $z(p)$ at t. We remove all points of $\mathfrak{R}$ for which $|z(p)| < \rho$. This leads to a Riemann surface $\mathfrak{R}^*$ with boundary and the variation for the abelian integrals and their periods can be given.

(b) We choose two points t_o and t_1 on $\mathfrak{R}$ with two uniformizers $z(p)$ and $\zeta(p)$ at t_o and t_1, respectively. We identify the circles $|z| = \rho$ and $|\zeta| = \rho$ by the correspondence: $z\,\zeta = e^{2i\alpha}$. This leads to a new Riemann surface $\mathfrak{R}^*$ which has one handle more than the original one. We can again express all functionals on $\mathfrak{R}^*$ in terms of these on $\mathfrak{R}$.

We do not give the explicit formulas but mention one distinctive fact for these topology changing variations. While a topology preserving variation (3.1) applied on a circle of radius ρ leads to a change of all functionals of order $O(\rho^2)$, the corresponding topology destroying variations lead to changes of order $O(-\frac{1}{\log \rho})$.

The main application of the topology destroying variations has been the construction of Riemann surfaces with prescribed topology from the sphere. We obtain asymptotic formulas for such Riemann surfaces which have prescribed structure but are infinitesimally near to the sphere whose functionals are particularly simple. These surfaces may be considered as elementary models of more complicated surfaces. Already Poincaré used such surfaces in order to study the relations between the various periods $P_{\mu\nu}$ and gave asymptotic formulas. Recently Garabedian used variational methods in order to investigate related problems.

6. Variations preserving conformal type

We may ask also for variations of $\mathfrak{R}$ which lead to a conformally equivalent surface $\mathfrak{R}^*$ that is to such a surface upon which $\mathfrak{R}$ may be mapped conformally. Such variations may be obtained by superimposing a sufficient number of elementary transformations (3.1) with proper choice of constants. We obtain easily the conditions for the conformal equivalence as follows. We choose a combination of integrals of the first kind

$$U = \sum_{\nu=1}^{G} \lambda_\nu U_\nu(p)$$

with real coefficients λ_ν whose derivative $U'(p)$ has $G-1$ different zeros on $\mathfrak{R}$. A variation $\mathfrak{R} \to \mathfrak{R}^*$ which is a combination of elementary variations of the type (3.1) leads to a surface $\mathfrak{R}^*$ whose points lie on $\mathfrak{R}$ but are identified in a different way. The integral of the first kind $U(p)$ goes over into the integral of the first kind $U^*(p)$ on $\mathfrak{R}^*$ which can be calculated by means of (3.3); $U^*(p)$ has also zero periods with respect to the odd cycles on $\mathfrak{R}^*$ which correspond to the odd cycles of the original surface $\mathfrak{R}$. If $\mathfrak{R}^*$ is conformally equivalent to $\mathfrak{R}$, there exists a one-to-one relation $p^* = p^*(p)$ between the two surfaces which realizes the conformal mapping. Since the integrals of the first kind go over into integrals of the first kind under such mappings, we have the identity $U^*(p^*) = U(p)$ which may be used in order to give an explicit expression for the relation $p^*(p)$. In order that such an equation be possible and have a single-valued solution, the values of U at the roots of $U'(p)$ must equal (up to an additive constant) to the values of U^* at the root of $U^{*\prime}(p)$; furthermore, the periods of $U(p)$ must be equal to the periods of $U^*(p)$ along corresponding cycles. It is easily seen that these conditions are necessary and sufficient for the conformal equivalence of $\mathfrak{R}$ and $\mathfrak{R}^*$ and they can easily be expressed in terms of the parameters $\varphi_\nu, \rho_\nu, t_\nu$ ($\nu = 1,\dots,N$) appearing in the combination of variations (3.1). It is true that the conditions in the variational formulas are fulfilled up to an order $o(\rho^2)$ only; but a standard argument in implicit functions shows that if the $o(\rho^2)$- terms are in the correct relation, a permissible correction can be found to satisfy the conditions exactly.

From the relation $U^*(p^*) = U(p)$ and from the variational formula (3.3) we can obtain an explicit formula for the conformal mapping of U onto U^*. We describe the points of $\mathfrak{R}^*$ by the same coordinates p as we used in $\mathfrak{R}$. By our construction the two surfaces differ only by the identification or deletion of certain points of $\mathfrak{R}$. The conformal mapping of $\mathfrak{R}$ onto $\mathfrak{R}^*$ shifts the point with the coordinate p into a point with the coordinate $p^*(p)$. This shift may be expressed in any given uniformizer $z(p)$ as follows:

$$z^* = z + \rho^2 r_z(z) + o(\rho^2). \tag{6.1}$$

If we change from the uniformizer z to a uniformizer $w = f(z)$ the variation will look in w as

$$w^* = w + \rho^2 r_w(w) + o(\rho^2) = f(z + \rho^2 r_z(z) + o(\rho^2)) \tag{6.2}$$

$$= w + \rho^2 r_z(z) f'(z) + o(\rho^2).$$

Hence, under a change of uniformizers, we have the transformation law

$$\frac{r_z(z)}{dz} = \frac{r_w(w)}{dw} \tag{6.3}$$

that is, $r(z)$ transforms like a reciprocal differential. The study of reciprocal differentials is natural to variational theory and Teichmüller was led to them by his own researches on extremum problems as well as Spencer and I.

The theory of reciprocal differentials can be reduced to the study of a certain kernel, the variation kernel, which can be easily constructed in terms of the bilinear differential $L(p,q)$. It is dependent on two points p and q on the surface and will be denoted by $n(p,q)$. It is a reciprocal differential in p and a quadratic differential of q. It has a simple pole with residue 1 if $p = q$ but has also $3G-6$ fixed poles $p_\nu \in \mathfrak{R}$ whose residues $Q_\nu(q)$ form a basis for all finite quadratic differentials on $\mathfrak{R}$.

A combination of local variations on $\mathfrak{R}$ of the type (3.1) may always be expressed in the form

$$p^* = p + \sum_{\nu=1}^{N} \lambda_\nu \rho^2 n(p,t_\nu) + o(\rho^2) \tag{6.4}$$

and the necessary and sufficient condition on the λ_ν to give a type preserving variation is the regularity of the variational term, that is:

$$\sum_{\nu=1}^{N} \lambda_\nu Q_j(t_\nu) = 0 \quad j = 1,2,\dots,3G-6, \tag{6.5}$$

the cancelling of the poles at the fixed singularities p_j of $n(p,q)$.

The variation of an open Riemann surface $\mathfrak{R}$ can also be expressed in a form similar to (6.4). We close $\mathfrak{R}$ to a surface $\mathfrak{F}$ by adding its double $\tilde{\mathfrak{R}}$. We introduce then the variation kernel $n(p,q)$ of $\mathfrak{F}$ and consider the variation

$$p^* = p + \sum_{\nu=1}^{N} \lambda_\nu \rho^2 n(p,\tilde{t}_\nu) + \sum_{\nu=1}^{N} \overline{\lambda}_\nu \rho^2 n(p,t_\nu) + o(\rho^2). \tag{6.6}$$

This variation leads again to a symmetric closed surface $\mathfrak{F}^*$ and taking one-half of $\mathfrak{R}^*$ of the new surface we obtain a varied open surface which corresponds to $\mathfrak{R}$. If all fixed poles p_j of the right-hand side of (6.6) cancel, we obtain a type-preserving variation of $\mathfrak{R}$ into $\mathfrak{R}^*$.

7. Variation of an imbedding

The variations of a Riemann surface which preserve its type lead to nothing new from the abstract point of view in the Riemann surface theory. But consider now the case that another surface $\mathfrak{M}$ is imbedded in $\mathfrak{R}$. The variation (6.4) will preserve $\mathfrak{R}$ but will change $\mathfrak{M}$. Hence, we possess a calculus of variations for solving extremum problems with respect to imbeddings of one surface into another.

We may go further and try to combine elementary variations on $\mathfrak{R}$ which preserve at the same time the conformal type of $\mathfrak{R}$ and of $\mathfrak{M}$. Let $\mathfrak{R}^*$ and $\mathfrak{M}^*$ be the surfaces after the variation. Mapping back $\mathfrak{R}^*$ into $\mathfrak{R}$, $\mathfrak{M}^*$ will go into a surface $\mathfrak{M}^\Delta$ imbedded in $\mathfrak{R}$ which is conformally equivalent to $\mathfrak{M}$ but does not coincide with it, in general. In this way we can vary a surface imbedded in $\mathfrak{R}$ under preservation of its type.

Using the representation (6.6) for the variation of a Riemann surface with boundary, we can show that the condition for type preservation of $\mathfrak{M}$ under a variation (6.4) is the fulfillment of the equations

$$\sum_{\nu=1}^{N} \lambda_\nu \omega_k(t_\nu) = 0, \quad k = 1,2,\dots,3G_0-6, \; G_0 = \text{genus of } \mathfrak{M}, \tag{7.1}$$

where the $\omega_k(p)$ are a basis for all finite quadratic differentials on

$\mathfrak{M}$ which satisfy on the boundary of $\mathfrak{M}$ the differential equation

$$\mathfrak{Q}_k(p) \left(\frac{dp}{dt}\right)^2 = \text{real}, \quad t = \text{boundary uniformizer}. \tag{7.2}$$

The reason for condition (7.2) is that we are in reality dealing with finite quadratic differentials on $\mathfrak{M} + \widetilde{\mathfrak{M}}$ and that (7.2) guarantees the analytic continuation of $\mathfrak{Q}_k(p)$ into the doubled domain.

It can be shown that if (6.5) and (7.1) are fulfilled, the necessary and sufficient conditions for the conformal equivalence of $\mathfrak{R}$ with $\mathfrak{R}^*$ and of $\mathfrak{M}$ with $\mathfrak{M}^*$ are all fulfilled up to the order of $o(\rho^2)$. But if we want to show now that by an additional slight variation, we can fulfill the conditions exactly, we are led to the interesting condition that the $Q_j(p)$ and the $\mathfrak{Q}_k(p)$ be linearly independent. If this condition is fulfilled, we have a great freedom in varying $\mathfrak{M}$ relative to $\mathfrak{R}$. If, however, some linear relation between the finite quadratic differentials of $\mathfrak{M}$ and $\mathfrak{R}$ exists, we are hindered in varying $\mathfrak{M}$ with respect to $\mathfrak{R}$. We are led to the phenomenon that $\mathfrak{M}$ may be rigidly imbedded in $\mathfrak{R}$, that is, that no conformally equivalent surface $\mathfrak{M}^\Delta$ may be fitted into $\mathfrak{R}$. Since a necessary condition for the rigid imbedding of $\mathfrak{M}$ into $\mathfrak{R}$ is an equation

$$\mathfrak{Q}(p) = Q(p), \tag{7.3}$$

$\mathfrak{Q}(p)$ and $Q(p)$ being finite quadratic differentials on $\mathfrak{M}$ and $\mathfrak{R}$, we recognize by (7.2) that the relative boundary of $\mathfrak{M}$ in $\mathfrak{R}$ is an analytic curve in $\mathfrak{R}$ satisfying the differential equation

$$Q(p) \left(\frac{dp}{dt}\right)^2 = \text{real}, \tag{7.4}$$

where t is a boundary uniformizer of $\mathfrak{M}$ and $Q(p)$ is an everywhere finite quadratic differential on $\mathfrak{R}$.

In order to show an example of a rigid imbedding, let $\mathfrak{R}$ be the triply-connected plane domain inside a Jordan curve b_0 from which two subdomains bounded by the Jordan curves b_1 and b_2 have been removed. There exists a continuum Γ connecting b_1 and b_2 in $\mathfrak{R}$ such that the doubly-connected domain $\mathfrak{R} - \Gamma$ has a maximum modulus among all double-connected domains imbedded in $\mathfrak{R}$. Then, it can be seen that $\mathfrak{R} - \Gamma$ is rigidly imbedded in $\mathfrak{R}$ since every variation of the domain will decrease its modulus and change, therefore, its conformal type.

8. Extremum problems in the theory of imbedding

The preceding example is already an illustration for a general type of extremum problems in imbedding. We may ask to imbed a doubly-connected planar domain $\mathfrak{M}$ into a given surface $\mathfrak{R}$ with or without boundary such that its modulus be maximum, or put analogous problems for other types of domains $\mathfrak{M}$.

We may give a fixed point t in $\mathfrak{R}$ and a fixed uniformizer at t, say $z(t)$. We may consider the Green's functions of all possible domains $\mathfrak{M}$ imbedded in $\mathfrak{R}$ which contain t and develop them near t in the form

$$G(p,t) = \log \frac{1}{|z|} + \gamma(t) + O(|z|). \tag{8.1}$$

$\gamma(t)$ is called the capacity of the domain $\mathfrak{M}$ at the point t in the uniformizer $z(t)$. We may consider $\gamma(t)$ as a measure of $\mathfrak{M}$ and compare various domains $\mathfrak{M}$ imbedded in $\mathfrak{R}$ by their capacity.

We are guaranteed by compactness arguments that an extremum domain must exist and we can compare its capacity with that of near-by imbedded comparison domains by means of the variational formula for the Green's function. The condition that $\gamma(t)$ decreases for every permissible variation leads to a linear dependence between the quadratic differentials on $\mathfrak{M}$ (which describe the change of $\gamma(t)$) and the quadratic differentials on $\mathfrak{R}$ (which appear because of the side conditions on preservation of conformal type and on keeping the point t and its uniformizer fixed under the variation). Thus, the boundaries of the extremum domains $\mathfrak{M}$ appear always as analytic curves on $\mathfrak{R}$ satisfying differential equations of the type

$$Q(p) \left(\frac{dp}{dt}\right)^2 = \text{real} \tag{8.2}$$

where $Q(p)$ is a quadratic differential on $\mathfrak{R}$ (not necessarily finite everywhere). It can also be shown by variational arguments that $\mathfrak{M}$ has no exterior points on $\mathfrak{R}$, that is, $\mathfrak{M}$ is obtained from $\mathfrak{R}$ by slitting the original surface along analytic arcs with the differential equation (8.2).

As a first application let us ask for a disk $\mathfrak{M}$ imbedded in $\mathfrak{R}$ which contains t and has a maximal capacity $\gamma(t)$. The above method leads to a differential equation for the boundary slits of the extremum domain $\mathfrak{M}$ which may be interpreted as follows. If $\mathfrak{M}$ is mapped upon the unit circle such that t corresponds to its center, each boundary arc of $\mathfrak{M}$ becomes an arc on the unit circumference. The two edges of each subarc of a boundary slit go over into two arcs of equal length on the unit periphery. Thus, the unit circle may be considered as a realization of the abstract surface $\mathfrak{R}$ if we identify isometrically corresponding arcs of its periphery. The possibility of such uniformization of each closed Riemann surface was first pointed out by Klein. We gave here an existence proof for the Klein mapping by using an extremum problem for which the existence of a solution is sure a priori.

We may construct the explicit solution of this extremum problem by a procedure which is applicable in many analogous cases. We cut $\mathfrak{R}$ by an arbitrary canonical cut system into a disk; on the other hand, we consider its universal covering surface $\mathfrak{W}$. All points on $\mathfrak{W}$ are identified if they lie over the same point $p \in \mathfrak{R}$ and can be connected with one fixed point $p_1 \in \mathfrak{W}$ over p by a curve in $\mathfrak{W}$ which cuts the canonical cut system an even number of times. The abstract point thus obtained is again denoted by the letter p. All points in $\mathfrak{W}$ over $p \in \mathfrak{R}$ which can be connected with p_1 by a curve cutting the canonical system an odd number of times form the associated point $\hat{p}$. We define thus a two-sheeted covering surface $\mathfrak{R} + \hat{\mathfrak{R}}$ of $\mathfrak{R}$ and let $V(p, p_0; q, q_0)$ be the single-valued real part of an abelian integral of the third kind on the doubled surface which is symmetric in its pairs of argument points. Then $1/2V(p, \hat{p}; t, \hat{t})$ is the Green's function of the extremum domain $\mathfrak{M}$ and the relative boundary of $\mathfrak{M}$ on $\mathfrak{R}$ is given by the equation

$$V(p, \hat{p}; t, \hat{t}) = 0. \tag{8.3}$$

Since the extremum domains in many imbedding problems are characterized by boundary curves satisfying quadratic differential equations, the introduction of a two-sheeted covering allows in many cases to express the solution of the problem in terms of the functionals of the doubled surface. Another aspect of the above solution is the possibility to characterize certain cross-cuts on a given surface with prescribed topological characteristics in a more specific way by extremum properties.

We illustrate this fact by the following example. Let Γ be a Jordan curve on $\mathfrak{R}$ which does not bound. We ask for curves b on $\mathfrak{R}$ which are homotopic to Γ and such that the capacity of the domain $\mathfrak{R} - \Gamma$ at the given point t and for a given uniformizer $z(t)$ be a maximum. The variational treatment leads to the following answer. Identify all points of the universal covering $\mathfrak{W}$ of $\mathfrak{R}$ if they lie over p and can be connected with one fixed replica p_1 by a curve which cuts Γ an even number of times. All other points of $\mathfrak{W}$ over p form the associated point $\hat{p}$. Consider the two-sheeted covering $\mathfrak{R} + \hat{\mathfrak{R}}$ of $\mathfrak{R}$. The extremum curve b is determined by the equation

$$\mathfrak{V}(p, \hat{p}; t, \hat{t}) = 0 \tag{8.4}$$

where $\mathfrak{V}$ is now the single-valued real part of the integral of the third kind with respect to the new two-sheeted covering of $\mathfrak{R}$. We may extend this type of extremum problems further and connect with each homotopy class of curves one canonical representative distinguished by an extremum property.

Consider finally the following problem. Two points t_0 and t_1

are given on the Riemann surface $\mathfrak{R}$ with prescribed local uniformizers $\zeta_o(p)$ and $\zeta_1(p)$; try to find two domains $\mathfrak{M}_o$ and $\mathfrak{M}_1$ on $\mathfrak{R}$ without common points such that $t_o \varepsilon \mathfrak{M}_o$ and $t_1 \varepsilon \mathfrak{M}_1$ and that $\gamma_o(t_o) + \gamma_1(t_1)$ be maximum where $\gamma_1(t_1)$ denotes the capacity of the domain $\mathfrak{M}_1$ at the point t_1. The answer to this question as given by the variational technique is the following. Let $V(p,p_o; q,q_o)$ be the single-valued real part of the integral of the third kind of $\mathfrak{R}$; the locus

$$V(p,p_o; t_o,t_1) = \text{const} \tag{8.5}$$

will subdivide $\mathfrak{R}$ into two domains $\mathfrak{M}_o$ and $\mathfrak{M}_1$. The two domains $\mathfrak{M}_o$ and $\mathfrak{M}_1$ obtained for any choice of the constant will be extremum domains and lead to the same value of the maximum.

In the case that $\mathfrak{R}$ is the sphere we may realize it over the complex plane and put

$$V(p,p_o; t_o,t_1) = \log \left| \frac{z - \xi_1}{z - \xi_o} \right| \tag{8.6}$$

by letting $t_o = \infty$. Thus, the extremum domains appear as circles in the complex plane and we find that

$$\gamma_o(\zeta_o) + \gamma_1(\xi_1) \leq 2 \log \left| \xi_1 - \xi_o \right| \tag{8.7}$$

for any two domains without common points containing ξ_o and ξ_1, respectively. This result is equivalent to an interesting theorem of Lavrentieff on pairs of functions $f(z)$ and $g(z)$ which are schlicht and meromorphic in the unit circle and have no common image points.

We focused the attention to the case of closed Riemann surfaces. We may formulate extremum problems of the same form for surfaces with boundary and solve them by the same methods. For example, the decomposition of a domain $\mathfrak{R}$ with boundary into two subdomains $\mathfrak{M}_o$ and $\mathfrak{M}_1$ with a maximum sum for the capacities at two points $t_i \varepsilon \mathfrak{M}_i$ is executed by the line

$$G(p,t_o) = G(p,t_1) \tag{8.8}$$

where G is the Green's function of $\mathfrak{R}$ and we have the inequality

$$\gamma_o(t_o) + \gamma_1(t_1) \leq \gamma(t_o) + \gamma(t_1) - 2G(t_o,t_1) \tag{8.9}$$

where $\gamma(t_1)$ is the capacity of $\mathfrak{R}$ at the point t_1 in the prescribed uniformizers.

9. Imbedding and schlicht mapping

Let $\mathfrak{R}$ be a Riemann surface containing the point π_0 with a local uniformizer $z(\pi)$. Let $\mathfrak{N}$ be another Riemann surface and let p_0, $w(p)$ be a point and a local uniformizer on it. Suppose that a mapping $p(\pi)$ of $\mathfrak{R}$ into a subdomain $\mathfrak{M} \subset \mathfrak{N}$ is possible such that $p_0 = p(\pi_0)$. In local uniformizers, it will have a series development at π_0 of the form

$$w = a_1 z + a_2 z^2 + \dots + a_n z^n \dots \quad . \tag{9.1}$$

If $\mathfrak{M}$ is not rigidly imbedded in $\mathfrak{N}$ there will exist an infinity of other mappings of $\mathfrak{R}$ into $\mathfrak{N}$. Using the variational theory for domains $\mathfrak{M}$ imbedded in $\mathfrak{N}$ of Section 7, we can easily show that to each mapping of $p = f(\pi)$ of $\mathfrak{R}$ into $\mathfrak{N}$ we have a nearby mapping $g^* = f^*(\pi)$ which transforms $\mathfrak{R}$ into a new subdomain $\mathfrak{M}^*$ of $\mathfrak{N}$. Let us assume for the sake of generality that $\mathfrak{R}$ and $\mathfrak{N}$ have both boundaries. We introduce the closed surfaces $\mathfrak{R} + \tilde{\mathfrak{R}}$ and $\mathfrak{N} + \tilde{\mathfrak{N}}$ and denote their variation kernels by $n(\pi, \kappa)$ and $N(p,q)$, respectively. It can be shown that the varied mapping functions $f^*(\pi)$ may be expressed in the form

$$f^*(\pi) = f(\pi) + \rho^2 [f'(\ ) \sum_{\nu=1}^{N} (\lambda_\nu n(\pi, t_\nu) + \bar{\lambda}_\nu n(\pi, \tilde{t}_\nu))$$

$$- \sum_{\nu=1}^{N} (\lambda_\nu f'(t_\nu)^2 N(f(\pi), f(t_\nu)) + \bar{\lambda}_\nu \overline{f'(t_\nu)^2} N(f(\pi), \widetilde{f(t_\nu)}))] + o(\rho^2). \tag{9.2}$$

The constants λ_ν have to satisfy the requirements

$$\sum_{\nu=1}^{N} \lambda_\nu \mathfrak{q}(t_\nu) = 0, \qquad \sum_{\nu=1}^{N} \lambda_\nu Q(f(t_\nu)) f'(t_\nu)^2 = 0 \tag{9.3}$$

for all finite quadratic differentials $\mathfrak{q}(\pi)$ on $\mathfrak{R}$ and $Q(p)$ on $\mathfrak{N}$ which satisfy on the boundaries of their respective domains of definition the differential equations (t = boundary uniformizer):

$$\mathfrak{q}(\pi) \left(\frac{d\pi}{dt}\right)^2 = \text{real}, \qquad Q(p) \left(\frac{dp}{dt}\right)^2 = \text{real}. \tag{9.3)'$$

If $\mathfrak{N}$ is a closed Riemann surface, we have to use in (9.2) the variation kernel $N(p,q)$ of $\mathfrak{N}$ itself and to omit the terms $N(f(\pi), \widetilde{f(t_\nu)})$. On the other hand, the second condition (9.3)' becomes vacuous and all finite quadratic differentials on $\mathfrak{N}$ have to be considered in the second condition (9.3). One recognizes easily that under these assumptions $f^*(\pi)$ gives an imbedding of $\mathfrak{R}$ into $\mathfrak{N}$ just as well as $f(\pi)$. Thus, we may characterize extremum functions for the imbedding problem by comparison with slightly varied competitors.

We can deal, in particular, with the coefficient problem, that is, determine the set of complex vectors $(a_1, a_2, \ldots, a_n)$ in n-space which can be the set of the n first coefficients in the development of an imbedding function (9.1). It can be shown that these vectors determine a bounded closed region of the complex n-space whose boundary satisfies a partial differential equation of the first order. A deeper study of this surface can be made by using the theory of characteristics which create the boundary surface. One is led to a generalization of Löwner's theory developed in the case of schlicht functions in a circle and obtains analogously ordinary differential equations for a one-parameter family of imbedding functions.

This general theory contains as particular applications the theory of schlicht functions in simply- and multiply-connected domains, the theory of bounded schlicht functions, the theory of p-valued functions and the theory of functions which do not accept a prescribed set of values. The abstract unified treatment of all imbedding problems does not lead to any additional complication and clarifies considerably, on the other hand, the general mechanism of the variational method applied.

The method becomes particularly simple if the domain $\mathfrak{N}$ is the sphere. Since the extremum domain $\mathfrak{M}$ in $\mathfrak{N}$ which is obtained from $\mathfrak{R}$ by the extremum mapping function $f(\mathfrak{m})$ is bounded by analytic arcs with the differential equation

$$(9.4) \qquad Q(p)\left(\frac{dp}{dt}\right)^2 = \text{real}, \quad Q(p) = \text{quadratic differential on } \mathfrak{N},$$

and since all quadratic differentials on $\mathfrak{N}$ are elementary and well-known, we can easily characterize these extremum domains. The nature of the original domain $\mathfrak{R}$ is quite unessential in this study.

Let, for example, $\mathfrak{R}$ be an m-times connected domain in the z-plane and consider all functions

$$(9.5) \qquad w = f(z) = z + a_2 z^2 + \ldots + a_n z^n + \ldots$$

which are regular and schlicht in $\mathfrak{R}$. Every function $f(z)$ of this class which solves some extremum problem involving the n first coefficients, maps on a slit domain $\mathfrak{M}$ in the w-plane whose boundary slits satisfy a differential equation:

$$(9.6) \qquad P_n\left(\frac{1}{w}\right)\left(\frac{dw}{dt}\right)^2 + 1 = 0$$

when $P_n(x)$ is a polynomial of degree $n + 1$. Conversely, all domains $\mathfrak{M}$ bounded by slits satisfying differential equations of the type (9.6) and which are conformally equivalent to $\mathfrak{R}$ belong to mapping functions for which the vector $(a_2,\dots,a_n)$ lies on the boundary of the coefficient region. It is interesting to remark that each characteristic curve on the boundary of the coefficient region with respect to the partial differential equation satisfied by this hypersurface determines an infinity of schlicht functions on $\mathfrak{R}$ such that the boundary slits of the corresponding image domains $\mathfrak{M}$ satisfy the same fixed differential equation (9.6). Thus, the set of characteristics stands in a very simple relation to the set of polynomials $P_n(x)$ of order $(n+1)$.

In a similar way, we may treat extremum problems with respect to any differentiable functional $\phi[f]$ which is bounded for the class (9.5) of schlicht functions in $\mathfrak{R}$.

10. Extremum problems for Riemann surfaces in 3-space

We have seen in Section 4 that the variational method may be used to investigate intrinsic properties of a given surface. But most of the important applications of variational calculus deal with non-intrinsic properties as imbedding, realization in the complex plane and dismemberment of surfaces. It is, therefore, natural to look for other realizations of Riemann surfaces and to study extremum problems and distortion theory there.

Let Σ be a surface in 3-space which serves as a realization of an abstract Riemann surface $\mathfrak{R}$. Let $V(p,p_o;\ q,q_o)$ denote the single-valued real part of the integral of the third kind on $\mathfrak{R}$. We may refer V to the local coordinates (u_1,u_2) of the surface Σ and express its properties there. If

$$ds^2 = \sum_{i,k=1}^{2} g_{i_k}\, d\, u_i\, d\, u_k \tag{10.1}$$

is the representation of the euclidean distance element ds in the coordinates in Σ, V satisfies as a function of each argument point the Beltrami equation

$$\sum_{i,k=1}^{2} \frac{\partial}{\partial \omega_k} \left(\sqrt{g}\ g^{ik}\ \frac{\partial V}{\partial u_i}\right) = 0 . \tag{10.2}$$

Let us perform now a deformation of the surface Σ and ask for the change of the "Green's function" V. We refer the deformed surface Σ^* to the same parameters (u_1,u_2) of Σ; however, since the geometry of Σ is now changed, we must take notice of the variation $\delta_g{}^{ik} = g^{*ik} - g^{ik}$. By using Green's identity, we are then led to the following variational formula:

$$(10.3) \qquad V^*(p,p_0;\, q,q_0) - V(p,p_0;\, q,q_0)$$

$$= -\iint \sum_{i,k=1}^{2} \delta(g^{ik}\sqrt{g}\,) \frac{\partial V(t,t_0;p,p_0)}{\partial u_i} \frac{\partial V(t,t_0;q,q_0)}{\partial u_k}\, du_1\, du_2 + o(\delta g^{ik}).$$

Here t is the point on $\mathfrak{R}$ which belongs to the coordinates (u_1,u_2) on Σ and t_0 is an arbitrary point on Σ. This formula leads immediately to corresponding variational results for the integrals of the first and second kind and for their periods. We can determine deformations of Σ which preserve its conformal type and deal with extremum problems relating the three-dimensional geometry of Σ to the conformal moduli of the abstract surface $\mathfrak{R}$. While the formal part of this variational theory is very similar to the corresponding plane case, the latter case can be treated with greater ease because of the compactness properties of families of schlicht mappings in the complex plane. These properties ensure a priori the existence of extremum domains in the plane case while in the problems considered in this section the existence proof for an extremum surface represents the more difficult part of the investigation.

REFERENCES

[1] BERGMAN, S., The kernel function and conformal mapping, New York, 1950.

[2] COURANT, R., Dirichlet's principle, conformal mapping and minimal surfaces, New York, 1950.

[3] GARABEDIAN, P. R., Asymptotic identities among periods of integrals of the first kind, Amer. J. Math., vol. 73 (1951), pp. 107-121.

[4] GARABEDIAN, P. R., SCHIFFER, M., Identities in the theory of conformal mapping, Transactions Amer. Math. Soc., vol. 65 (1948), pp. 187-238.

[5] GOLUSIN, G., Method of variation in the theory of conformal mapping, Receuil Math. (Math. Sbornik), N.S. Vol. 19 (1946), pp. 203-236; vol. 21 (1947), pp. 83-117, 119-132; vol. 29 (1951), pp. 455-468.

[6] LÖWNER, K., Untersuchungen über schlichte konforme Abbildungen des Einheitskreises, Math. Annalen, vol. 89, (1923), pp. 103-121.

[7] SCHAEFFER, A. C., SPENCER, D. C., Coefficient regions for schlicht functions, Amer. Math. Soc. Colloquium Publications, vol. 35, New York, 1950.

[8] SCHIFFER, M., SPENCER, D. C., The coefficient problem for multiply-connected domains, Ann. of Math., vol. 52 (1950), pp. 362-402.

[9] SCHIFFER, M., SPENCER, D. C., A variational calculus for Riemann surfaces, Ann. Acad. Scient. Fennicae, A. I. 93.

[10] SCHIFFER, M., SPENCER, D. C., On the conformal mapping of one Riemann surface into another, Ann. Acad. Scient. Fennicae, A. I. 94.

[11] SCHIFFER, M., SPENCER, D. C., Functionals of finite Riemann surfaces, Princeton University Press, to appear.

[12] TEICHMÜLLER, O., Extremale quasikonforme Abbildungen und quadratische Differentiale, Abh. preuss. Akad. Wiss., Math.-nat. Kl. 1940, pp. 1-197.

SEMIGROUPS OF TRANSFORMATIONS OF A RIEMANN SURFACE INTO ITSELF

PAUL C. ROSENBLOOM

Introduction

In the theory of Lie groups one shows that local coordinate systems can be introduced in which the group operation is analytic. Comparatively little has been done, however, in working out the implications of the deeper properties of analytic functions in relation to the structure of Lie groups. This paper is a small contribution in this direction.

We consider a one parameter family of analytic transformations of a Riemann surface S_1 into itself which is closed under composition. If $f(z_1,z_2)$ is the point into which z_1 is mapped by the transformation corresponding to the parameter z_2, then f must satisfy the functional equation

$$f(f(z_1,z_2),z_3) = f(z_1,g(z_2,z_3)), \tag{1}$$

where $g(z_2,z_3)$ is the function expressing the dependence of the product of two transformations in the family on the factors. We shall suppose that the domain of the parameter is a Riemann surface S_2, that f is analytic on $S_1 \times S_2$, and that g is analytic on $S_2 \times S_2$. The fundamental problem is to determine all pairs of functions f and g which satisfy (1). We shall solve this problem in the case where S_1 and S_2 are parabolic. By the uniformization theorem, we can assume that S_1 and S_2 are the complex plane, so that the problem reduces to that of the determination of all pairs, f and g, of entire functions which satisfy (1).

If we set $f = g$, then we obtain the functional equation

$$f(f(z_1,z_2),z_3) = f(z_1,f(z_2,z_3)), \tag{2}$$

which says that the surface S_1 is an analytic semigroup with respect to the operation f on $S_1 \times S_1$ to S_1. We prove that the only entire functions satisfying (2) are

$$f(z_1,z_2) = c,\ z_1,\ z_2,\ z_1 + z_2 - c, \tag{3}$$

and

$$a(z_1 - b)(z_2 - b) + b,$$

where a, b, and c are constants and $a \neq 0$. There are six classes of pairs, f and g, of entire functions which satisfy (1):

(1a) $f = c$ or $f = z_1$, g arbitrary;

(1b) $g(z_1,z_2) = z_2 + c$, $f = \phi(e^{2\pi i z 2/c})$, $\phi(\omega)$ analytic for $\omega \neq 0$;

(1c) $g = b(z_2-c) + c$, $b^n = 1$, $f = \phi((z_2-c)^n)$;

(1d) $g = a(z_1-b)(z_2-b) + b$, $f = (z_1-b)(a(z_2-b))^n + b$;

(1e) $g = z_1 + z_2 - a$, $f = z_1 + b\ (z_2-a)$;

(1f) $g = z_1 + z_2 - a$, $f = e^{c(z_2-a)}(z_1-b) + b$;

where a, b, and c are arbitrary constants.

The method is based on Picard's theorem and an easy extension which is a special case of a theorem of Borel [3] to the effect that if $F(z)$ is an entire function and if $F - z$ and $F - a$ have at most one zero, then F is a linear function. This follows immediately from the fact that if the meromorphic function $(F-z)/(F-a)$ takes on each of the values 0, 1, and ∞ at most once, then it must be a rational function of the first degree.

In the course of this investigation, it is necessary to treat also the functional equations

$$f(g(z)) = f(z+A), \tag{4}$$

and

$$F(f(f(z_1,z_2),z_3)) = F(f(z_1,f(z_2,z_3))). \tag{5}$$

The solutions of (4) are:

(4a) $f = c$, g arbitrary,

(4b) $g = z + A$, f arbitrary,

(4c) $g = z + c$, $f = \phi(e^{2\pi i z/(c-A)})$, $c \neq A$, $\phi(\omega)$ analytic for $\omega \neq 0$,

(4d) $g = a(z-b) + b$, $a^n = 1$, $f = \phi((z-b+aA/(1-a))^n)$.

The solutions of (5) are:

(5a) $F = c$, f arbitrary,

(5b) f as in (3), F arbitrary,

(5c) $f(z_1,z_2) = a(z_1-b) + b$, $a^n = 1$, $F = \phi((z-b)^n)$,

(5d) $f(z_1,z_2) = a(z_2-b) + b, \quad a^n = 1, \quad F = \phi((z-b)^n),$

(5e) $f(z_1,z_2) = z_1 + c$ or $z_2 + c, \quad F = \phi(e^{2\pi i z/c}).$

The solution of (4) is a trivial consequence of a theorem of Bohr [2], which we use in the precise form given by Hayman [4].

Related functional equations, in particular (2), have been studied by Abel [1] and Stäckel [6] by completely different methods. It is not quite clear how their methods would have to be modified in order to meet modern standards of rigor.

We might mention that we have so far not succeeded in treating the "distributive" law:

$$f(g(z_1,z_2)) = g(f(z_1),f(z_2))$$

by the present methods.

The results on equation (2) have been presented to the American Mathematical Society. [5]

Known Lemmas

LEMMA 1. If $f(z)$ is an entire function, then either $f - z$ or $f - a$ has a zero. For if not, then f must be linear, which immediately leads to a contradiction.

LEMMA 2. If $g(z)$ is analytic for $|z| < r$ and $R(r,g)$ is the least upper bound of the radii σ such that the whole circumference $|\omega| = \sigma$ is contained in the image of $|z| < r$ by the mapping $\omega = g(z)$, then

$$M(\rho,g) \leq 4\rho r/(r-\rho)^2 \, (R(r,g) + |g(0)|) + |g(0)|.$$

for

$$0 \leq \rho < r.$$

Here, as in the future, we shall denote by $M(\rho,g)$ the least upper bound of $|g(z)|$ for $|z| \leq \rho$. This is the best bound, found by Hayman, for $R(r,g)$ in terms of $M(\rho,g)$; the existence of such a bound had been proved by Bohr.

COROLLARY. If f is an entire function and g is analytic for $|z| < r$, then

$$M(r,f(g(z))) \geq M(1/8(M(r/2,g) - |g(0)|) - |g(0)|,f).$$

Equation (4)

THEOREM 1. The solutions of (4) are given by (4a) - (4d). We may assume that F is not constant.

PROOF. Let $g_1(z) = g(z) - g(0)$, so that $g_1(0) = 0$, and set

$$F_1(z) = F(z+g(0)).$$

Hence

$$F_1(g_1(z)) = F_1(z+A_1),$$

where $A_1 = A - g(0)$. Then

$$M(1/8\ M(r,g_1),F_1) \leq M(2r,F_1(g_1(z)))$$
$$\leq M(2r+|A_1|,F_1),$$

and consequently

$$M(r,g_1) \leq 8(2r+|A_1|),$$

which implies that $g_1(z) = az$, where a is constant. Thus

$$M(r-|A_1|,F_1) \leq M(|a|r,F_1) \leq M(r+|A_1|,F_1),$$

so that $|a| = 1$. If $a \neq 1$, let

$$F_2(z) = F_1(\beta+z), \quad \beta = aA/(a-1).$$

Then $F_2(az) = F_2(z)$. Hence a is an nth root of unity for some $n > 1$, and $F_2(z) = \phi(z^n)$, where ϕ is an entire function. The theorem immediately follows.

COROLLARY 1a. If equation (5) holds and if f is a function of z_1 or z_2 alone, then one of (5a) - (5e) holds.

Equations (2) and (5)

It is convenient to treat these together. For any a there is, by Lemma 1, a b such that $f(b,a) = a$ or $f(b,a) = b$. By symmetry, we may concentrate on the latter case. We shall assume that F is not constant and that f is not a function of z_1 or z_2 alone, for this case is already taken care of by Corollary 1a.

LEMMA 3. If (5) and $f(b,a) = b$, then either $f(z,b) \equiv$ constant or $F(f(z,a)) \equiv F(z)$.

PROOF. We have $F(f(f(z,b),a) = F(f(z,b))$. Now if $f(z,b)$ is not constant, then its range contains an open set U, and for $z \in U$ we have $F(f(z,a) = F(z)$, and hence this equation holds identically.

LEMMA 4. If $\phi(\psi(z_1,z_2)) = \chi(z_1)$ and ϕ, ψ and χ are entire functions, then either ϕ is constant or ψ is a function of z_1 alone. A similar conclusion holds if χ is a function of z_2 alone.

LEMMA 5. If $\phi(z_1,\psi(z_2)) = \chi(z_2)$ and ϕ, ψ, and χ are entire functions, then either ψ and χ are constant or ϕ is a function of z_2 alone.

These two lemmas are trivial.

LEMMA 6. If (5) and $f(z,b) \equiv c$, then $f(z,b) \equiv f(b,z) \equiv f(c,z) \equiv f(z,c) \equiv c$.

PROOF. We have $F(f(z_1,f(b,z_2))) = F(f(c,z_2))$. By Lemmas 4 and 5, since f is not a function of z_2 alone, then $f(b,z)$ and $f(c,z)$ are constant. Hence $f(b,z) = f(b,b) = c$ and $f(c,z) = f(c,b) = c$. The rest follows by symmetry.

COROLLARY 6a. Under the same hypotheses, if $f(\alpha,\beta) = c$, then either $f(z,\alpha) \equiv c$ or $f(z,\beta) \equiv c$.

COROLLARY 6b. Under the same hypotheses, if $f(\alpha,\beta) = \beta$, then either $F(f(\alpha,z)) \equiv F(z)$ or $\beta = c$.

Corollary 6a follows from the equation $F(f(f(z,\alpha),\beta)) = F(c)$, and Corollary 6b from Lemma 3.

LEMMA 7. If $F(f_1(z_1,\ldots,z_k)) \equiv F(f_2(z_1,\ldots,z_k))$ and $f_1(z_{1_0},\ldots,z_{ko}) = f_2(z_{1o},\ldots,z_{ko}) = w_o$ where $F'(w_o) \neq 0$, then $f_1 = f_2$.

COROLLARY 7a. If $F(f_1(z_1,\ldots,z_k)) \equiv F(f_2(z_1,\ldots,z_k)$ and F is not constant, and if for every w there is a point $(z_1,z_2,\ldots,z_k)$ such that $f_1(z_1,\ldots,z_k) = f_2(z_1,\ldots,z_k) = w$, then $f_1 \equiv f_2$.

COROLLARY 7b. If (5) and $f(z,a) = z$, then (2).

This lemma and its corollaries are trivial.

We shall now settle equation (2) under the assumption that $f(z,a) \equiv z$.

LEMMA 8. If (2) and $f(z,a) \equiv z$, then $F(a,z) \equiv z$.

PROOF. By hypothesis, $f(a,a) = a$. Hence, by Lemma 3, either $f(a,z)$ is constant or $f(a,z) \equiv z$. If the former, then $f(a,z) = f(a,a) = a$, and therefore $f(z_1,z_2) = f(f(z_1,a),z_2) = f(z_1,a) = z_1$, which has already been excluded.

LEMMA 9. If (2) and $f(z,a) = z$, and if $f(c,d) = c$, $d \neq a$, then

$$f(z_1,z_2) = (z_1-c)(z_2-c)/(a-c) + c.$$

PROOF. By Lemma 3, either $f(z,c)$ is constant or $f(z,d) \equiv z$. But $f(a,d) = d$, by Lemma 8, so that the latter is impossible. Hence $f(z,c) = f(a,c) = c = f(c,z)$, by Lemma 6. Now if $f(\alpha,\beta) = \beta$, then either $\beta = c$ or $\alpha = a$, by Corollary 6b. Consequently, if $\beta \neq c$, then $f(z,\beta) \neq \beta$ for $z \neq a$, and $f(z,\beta) \neq z$ for $z \neq c$. Therefore $f(z,\beta)$ is a linear function: $f(z,\beta) = A(\beta)\, z + B(\beta)$. Now $B(\beta) = f(0,\beta)$ and $A(\beta) = f(1,\beta) - B(\beta)$, so that the coefficients are entire functions. On solving the equation

$$z = f(a,z) = A(z)a + B(z), \quad c = A(z)b + B(z),$$

we obtain the above expression for f. Clearly the same result holds if $f(c,d) = d$ for some $c \neq a$.

LEMMA 10. If (2) and $f(z,a) \equiv z$, and if $f(z_1,z_2) \neq z_1$ for $z_2 \neq a$, and $f(z_1,z_2) \neq z_2$ for $z_1 \neq a$, then $f(z_1,z_2) = z_1 + z_2 - a$.

PROOF. If $\alpha \neq a$, then $f(z,\alpha) \neq \alpha$ for $z \neq a$ and $f(z,a) \neq z$ for all z. Hence $f(z,\alpha)$ is linear and must have the form $f(z,\alpha) = z + B(\alpha)$, where $B(\alpha) = f(0,\alpha)$ is an entire function. Therefor $B(\alpha) = f(a,\alpha) - a = \alpha - a$, which yields the desired conclusion.

It remains only to deal with the case where there is no a such that $f(z,a) \equiv z$ or $f(a,z) \equiv z$. We can, however, prove the more general Lemma 11.

LEMMA 11. If (5) and if there is no a such that $F(f(z,a)) \equiv F(z)$ or $F(f(a,z)) \equiv F(z)$, then f is constant.

PROOF. If $f(b,a) = b$, then $f(z,b)$ is constant, by Lemma 3, and so $f(b,z) = f(b,a) = b = f(z,b)$, by Lemma 6. If $\alpha \neq \beta$, then $f(\alpha,z) \neq z$ for $z \neq b$, by Corollary 6b, and $f(\alpha,z) \neq \alpha$ for all z. Hence $f(\alpha,z)$ is linear, and therefore constant, so that $f(\alpha,z) = f(\alpha,b) = b$, for all $\alpha \neq b$.

THEOREM 2. If (2), then f is one of the functions in (3).

This follows from Corollary 1a, Lemmas 3, 8, 9, 10 and 11.

Corollaries 1a and 7b and Lemma 11 also settle certain cases of Equation (5). To complete the treatment of (5), it suffices to assume that F is not constant, that f is not a function of z_1 or z_2 alone, and that there is no a such that $f(z,a) \equiv z$ or $f(a,z) \equiv z$.

LEMMA 12. If (5) and $f(b,a) = b$, then either $f(z,b) \equiv f(b,z) \equiv b$ or $F'(a) = F'(b) = 0$.

PROOF. By Lemma 3, either $f(z,b)$ is constant or $F(f(z,a)) \equiv F(z)$. By Lemma 6, if $f(z,b)$ is constant, then $f(z,b) \equiv f(b,z) \equiv f(b,a) = b$. Otherwise, $f(z,a) = A(z-b) + b$, where $A^n = 1$, and $F(w) = \phi((w-b)^n)$, by Theorem 1. Hence $F'(b) = 0$. Choose c so that $f(c,a) = a$. Then, by Lemma 3, either $f(a,z)$ is constant or $F(f(c,z)) \equiv F(z)$. But if $f(a,z)$ is constant, then so is $f(z,a)$, by Lemma 6, which leads to a contradiction. Hence, by Theorem 1 again, $f(c,z) = B(z-a) + a$, $B^k = 1$, and $F(w) = \phi((w-a)^k)$, which implies that $F'(a) = 0$.

LEMMA 13. If (5), then there is a c such that $f(c,z) = f(z,c) = c$.

PROOF. Let $F'(a) \neq 0$. By Lemma 1, there is a b such that $f(b,a) = b$ or $f(b,a) = a$. By Lemma 12, either $f(z,b)$ or $f(a,z)$ is constant, and the result follows from Lemma 6.

THEOREM 3. The solutions of (5) are given by (5a) - (5e).

PROOF. It remains only to consider the case where F is not constant, f is neither a function of z_1 nor of z_2 alone, there is no a such that $f(z,a) \equiv z$ or $f(a,z) \equiv z$, and there is a c such that $f(c,z) \equiv f(z,c) \equiv c$. Let $F'(a) \neq 0$, $a \neq c$. If $f(b,a) = b$, then $f(z,b) \equiv f(c,b) = c \equiv f(b,z) = f(b,a) = b$, by Lemmas 3 and 6. Hence $f(z,a) \neq z$ for $z \neq c$. Similarly $f(z,a) \neq a$ for all z. Therefore $f(z,a)$ is constant, in fact, $f(z,a) \equiv f(c,a) = c$. Consequently $f \equiv c$.

Equation (1)

From (1) we immediately obtain $f(z_1,g(g(z_2,z_3),z_4)) = f(z_1,g(z_2,g(z_3,z_4)))$, so that if $F(z) = f(z_1,z)$ for a fixed z_1, then (5) is satisfied by F and g. All the cases except (5b) are trivial and need not be discussed here, as are also the cases of (5b) where g is one of the first three functions in (3). By obvious changes in the coordinate system, it suffices to discuss the cases where $g(z_1,z_2) \equiv z_1z_2$ or $z_1 + z_2$, and f is not constant.

LEMMA 14. If $f(z)$ is a univalent entire function, then f is linear.

PROOF. This is a trivial consequence of the classical Weierstrass theorem. For if $|z| > 1$, then $f(z)$ omits the image of the unit circle by the mapping $w = f(z)$, which implies that f is a polynomial.

LEMMA 15. If (1) and $g(z_1,z_2) \equiv z_1z_2$, then $f(z_1,z_2) \equiv (z_1-c)\, z_2^n + c$, where n is an integer ≥ 0, and c is constant.

PROOF. We have $f(f(z_1,z_2),1) = f(z_1,z_2)$, so that $f(z,1) \equiv z$. If $z_2 \neq 0$, then $f(f(z_1,z_2),z_2^{-1}) = z_1$, so that for fixed z_2, $f(z,z_2)$ is a univalent function of z. Hence $f(z_1,z_2) = z_1\, A(z_2) + B(z_2)$, where A and B are entire functions. By (1), we infer that $A(z_1)A(z_2) = A(z_1z_2)$ and $B(z_1z_2) = B(z_1)\, A(z_2) + B(z_2)$. By a classical theorem of Cauchy, $A(z) = z^n$ for some integer n. Hence $B(z) = B(0)(1-z^n)$.

LEMMA 16. If (1) and $g(z_1,z_2) \equiv z_1 + z_2$, then $f(z_1,z_2) \equiv z_1 + bz_2$ or $e^{cz_2}(z_1-b) + b$, where b and c are constant.

PROOF. We have $f(f(z_1,z_2),0) = f(z_1,z_2)$, so that $f(z,0) = z$. Hence $f(f(z_1,z_2), -z_2) = z_1$, so that for fixed z_2, $f(z_1,z_2)$ is univalent. Thus, as before, $f(z_1,z_2) = z_1A(z_2) + B(z_2)$, where A and B are entire functions. As before, we obtain the equations $A(z_1+z_2) = A(z_1)A(z_2)$ and $B(z_1+z_2) = B(z_1)\, A(z_2) + B(z_2)$. Again, by Cauchy's theorem, $A(z) = e^{cz}$ where c is constant. But $B'(z_1+z_2) = B'(z_1)A(z_2)$, so that $B'(z) = B'(0)\, e^{cz}$. The above alternatives now follow according as

$c = 0$ or $c \neq 0$.

THEOREM 4. The solutions of (1) are given by (1a) - 1f).

BIBLIOGRAPHY

[1] ABEL, N. H., Oeuvres, ed. Sylow et Lie, vol. I, p. 61.

[2] BOHR, H., Über einen Satz von Edmund Landau, Scripta Univ. Hierosolymitanarum, vol. 1, nr. 2., (1923).

[3] BOREL, E., Sur les zéros des fonctions entières, Acta Math., vol. 20, (1897).

[4] HAYMAN, W. K., Symmetrization in the Theory of Functions, Stanford University, Technical Report no. 11, Office of Naval Research, (1950).

[5] ROSENBLOOM, P. C., Abstract, Bull. Amer. Math. Soc., vol. 52, p. 1009, (1946).

[6] STÄCKEL, P., Zeitschrift Math. Phys. vol. 42, (1897).

AN EXTREMAL BOUNDARY VALUE PROBLEM

A. C. SCHAEFFER

It is well known that if a function

$$f(z) = a_1 z + a_2 z^2 + \dots$$

is schlicht and regular in $|z| < 1$ then it has a radial limit almost everywhere on the circumference of the unit circle. In connection with an application of conformal mapping there arose the question: if E is a given closed set of positive measure on $|z| = 1$ is there a function $f(z)$ regular and schlicht in $|z| < 1$ such that $w = f(z)$ maps $|z| < 1$ onto $|w| < 1$ cut by radial slits, the set E mapping onto $|w| = 1$ and the complement of E mapping onto the radial slits. For the general class there is no such function, however, in the present note it is shown that each closed set E of positive measure has a perfect subset of measure equal to that of E for which the required mapping function $f(z)$ exists. The existence of such a function can be proved by the use of Green's function. However, Green's function is harmonic over the full plane outside of E except for its logarithmic singularity, while the definition of the function $f(z)$ concerns only the circle $|z| \leqq 1$. It is of interest to define $f(z)$ as an extremal schlicht function for this reason, and because there are analogous questions for the class of functions defined below.

Given a closed set E on $|z| = 1$ of positive measure let S_E be the class of functions

$$f(z) = a_1 z + a_2 z^2 + \dots \tag{1}$$

schlicht and regular in $|z| < 1$ such that

$$\lim_{r \to 1} f(re^{i\theta}) \geqq 1$$

for almost all $e^{i\theta} \in E$. It is to be shown that there is a function $f(z)$ of class S_E for which $|a_1|$ attains its minimum value, and that $w = f(z)$

* The results presented here represent work done under contract with the Office of Naval Research.

maps $|z| < 1$ onto $|w| < 1$ cut by radial slits. The complement of E maps into the slits and almost all points of E map into $|w| = 1$.

It is first to be noted that in each class S_E there is a positive lower bound to $|a_1|$. For if $f(z)$ is given by (1) then the function

$$\frac{za_1}{f(z)} = 1 + b_1 z + \ldots$$

is regular not zero in $|z| < 1$ and is bounded by 4 there, at almost all points of E it is bounded by $|a_1|$. Then the harmonic function $\log |z\, a_1/f(z)|$ has average value zero on the unit circumference, and hence

$$|E| \log |a_1| + (2\pi - |E|) \log 4 \geq 0.$$

This gives a lower bound for $|a_1|$.

There is then a function of class S_E for which a_1 attains its lower bound. For there is a sequence of functions

$$f_n(z) = a_1^{(n)} z + a_2^{(n)} z^2 + \ldots$$

which belongs to S_E for $n = 1,2,3,\ldots$ and converge in $|z| < 1$, with $|a_1^{(n)}| \to \inf |a_1|$. The limit function

$$f(z) = a_1 z + a_2 z^2 + \ldots$$

is regular and schlicht in $|z| < 1$. Since $z/f_n(z)$ is uniformly bounded its boundary values $e^{i\theta}/f_n(e^{i\theta})$ converge weakly in mean for some subsequence of n to a function which is bounded by 1 almost everywhere in E. Thus $e^{i\theta}/f(e^{i\theta})$ is bounded by 1 almost everywhere in E, and $f(z)$ belongs to S_E.

Thus for a given closed set E of positive measure let $f(z)$ be a function of class S_E which minimizes $|a_1|$. It is next to be shown that this function is bounded by 1 in $|z| < 1$. For suppose the contrary, then $|f(e^{i\theta})| > 1$ in a set E' of positive measure on $|z| = 1$. For $0 < r < 1$ define D_r as the image in the w plane of the circle $|z| < r$ under the mapping $w = f(z)$. The boundary of D_r is an analytic Jordan curve which for r sufficiently near 1 lies only partly in $|w| \leq 1$. The domain D_r consists of a domain D_r', the set of points in $|w| < 1$ that can be connected to $w = 0$ by a path lying in both D_r and $|w| < 1$, plus a finite set of lobes $L_1, L_2, \ldots, L_k$ lying at least partly in $|w| > 1$. The lobes are those points of D_r that cannot be connected to $w = 0$ by a path lying in both D_r and $|w| < 1$, and L_ν is joined to D_r' along an arc I_ν of the unit circumference $|w| = 1$. To the full unit circle $|w| < 1$, counted as sheet number zero, adjoin these lobes, each lobe being regarded as in a different sheet and adjoined to $|w| < 1$ along an arc I_ν of $|w| = 1$. A point in moving from $|w| < 1$ in the zero sheet across I_ν into the νth lobe moves into the νth sheet. The resulting multi-sheeted domain is simply connected, let it be Γ_r, and let it be mapped onto $|\zeta| < 1$ by

$$\zeta = \phi_r(w) = c_1(r)\, w + c_2(r)\, w^2 + \dots$$

The domains Γ_r form a nested sequence, increasing as r increases, and consequently $|c_1(r)|$ decreases. The function

$$\phi_r(f(z)) = c_1(r)a_1 z + \dots$$

is regular and schlicht in $|z| < r$ and is of modulus 1 wherever $|f(z)| \geqq 1$ on $|z| = r$. Under the supposition that $|f(e^{i\theta})| > 1$ in a set E' of positive measure, there is, according to Egeroff's theorem on uniform convergence, a set E'' of positive measure such that $|f(re^{i\theta})| \geqq 1$ for $e^{i\theta} \in E''$ and $r_0 \leqq r \leqq 1$. This shows that the modulus of $c_1(r)a_1$ has a positive lower bound as r approaches 1, and consequently $|c_1(r)|$ tends to a limit not zero as r approaches 1. Thus a selection theorem shows that a subsequence of the functions $\phi_r(w)$ approaches a limit

$$\phi_1(w) = c_1(1)\, w + c_2(1)\, w^2 + \dots \quad ,$$

and the nesting property then shows that the entire sequence approaches this limit as r approaches 1. Let D_1 be the outer limiting set of D_r, $0 < r < 1$. Under the mapping $\zeta = \phi_1(w)$ those accessible boundary points of D_1 that can be connected to $w = 0$ by a path lying in both D_1 and $|w| < 1$ map into points of $|\zeta| < 1$, the remaining accessible boundary points of D_1 map into $|\rho| = 1$. Thus

$$\lim_{r \to 1-} \phi_1\left(f(re^{i\phi})\right)$$

is equal to 1 for almost all points of E, and clearly $|c_1(1)a_1| < |a_1|$. This contradiction shows that the extremalizing function $f(z)$ is bounded by 1 in $|z| < 1$.

It is now to be shown that the image of $|z| < 1$ under $w = f(z)$ is $|w| < 1$ cut by radial slits. Let I be an arc of $|z| = 1$ complementary to E, and, neglecting sets of measure zero, increase I as much as possible in the complement of E. Thus almost all points of the arc I lie in the complement of E, but I does not lie in any larger arc with this property. If A is the image of I in $|w| \leqq 1$ then the part of $|w| < 1$ which is connected to $w = 0$ by a path in $|w| < 1$ not touching

A is a simply connected domain that can be mapped onto $|s| < 1$ by some function

$$s = g(w) = e_1 w + e_2 w^2 + \dots$$

Then $g(f(z))$ is regular, schlicht and bounded by 1 in $|z| < 1$, and its radial limit has modulus 1 at almost all points of the set $E + I$ on $|z| = 1$. Since almost all points of I map into points of $|s| = 1$ it follows that all points of I map into points of $|s| = 1$ and $g(f(z))$ can be continued analytically across I. Let J be the image of I in the s plane and rotate so that J is the arc $e^{i\phi}$, $-\phi_1 < \phi < \phi_1$. The arc CJ of $|s| = 1$ complementary to J is mapped by $w = g^{-1}(s)$ into an arc of $|w| = 1$. If

$$\psi(s) = h_1 s + h_2 s^2 + \dots$$

is regular and schlicht in $|s| < 1$ and $|\psi(s)| = 1$ on the arc CJ then

$$\psi(g(f(z))) = h_1 e_1 a_1 z + \dots$$

belongs to a class S_E. Thus

$$|h_1| \geq \frac{1}{|e_1|} \tag{2}$$

with equality if $\psi(s) = g^{-1}(s)$.

In order to obtain the minimizing function $\psi(s) = g^{-1}(s)$, a variational formula given by D. C. Spencer and the present author is used. Let Γ be an analytic Jordan arc in $|s| < 1$ with end-points α, β and let $p_\varepsilon(s)$, $|\varepsilon| \leq \varepsilon_0$, be a sequence of functions which are regular and uniformly bounded in a domain containing Γ and there satisfy

$$|p_{\varepsilon'}(s) - p_{\varepsilon''}(s)| \leq M\,|\varepsilon' - \varepsilon''|.$$

Suppose also that $p_\varepsilon(\alpha) = p_\varepsilon(\beta) = 0$, and let Γ_ε be the arc traced by $s + \varepsilon\, p_\varepsilon(s)$ when s traces Γ. Then [1], page 29, there is a mapping

$$s_1 = s\left\{1 + \frac{\varepsilon}{2\pi i}\int_\alpha^\beta \frac{p(u)}{2u^2}\,\frac{u+s}{u-s}\,du - \frac{\bar{\varepsilon}}{2\pi i}\int_\alpha^\beta \frac{\bar{p}(u)}{2\bar{u}^2}\,\frac{1+\bar{u}s}{1-\bar{u}s}\,d\bar{u} + o(\varepsilon)\right\},$$

of the one or two sheeted domain bounded by $|s| = 1$, Γ_ε, Γ onto $|s_1| < 1$ cut by a Jordan arc such that points $s \in \Gamma$, $s + \varepsilon\, p_\varepsilon(s) \in \Gamma_\varepsilon$ map into the same point of $|s_1| < 1$. Here $p(u) = p_0(u)$ is the value of $p_\varepsilon(u)$ corresponding to $\varepsilon = 0$. The length of the arc J is not, in general an invariant under this mapping, so let

$$s_2 = \frac{s_1 - \gamma\varepsilon}{1 - \bar{\varepsilon}\bar{\gamma} s_1}$$

be a bilinear transformation such that J maps into an arc of $|s_2| = 1$ of its own length. Then

$$s_2 = s\left\{1 + \frac{\varepsilon}{2\pi i}\int_\alpha^\beta \frac{p(u)}{2u^2}\,\frac{u+s}{u-s}\,du - \frac{\bar{\varepsilon}}{2\pi i}\int_\alpha^\beta \frac{\bar{p}(u)}{2\bar{u}^2}\,\frac{1+\bar{u}s}{1-\bar{u}s}\,d\bar{u}\right\} -$$

$$-\gamma\varepsilon + \bar{\gamma}\bar{\varepsilon}s^2 + o(\varepsilon)$$

where

$$\gamma = \frac{1}{e^{-i\phi_1} - e^{i\phi_1}} \cdot \frac{1}{2\pi i}\int_\alpha^\beta \frac{p(u)}{2u^2}\left\{\frac{u+e^{i\phi_1}}{u-e^{-i\phi_1}} - \frac{u+e^{i\phi_1}}{u-e^{i\phi_1}}\right\}du + o(1). \qquad (3)$$

Hence

$$s = s_2 - \frac{\varepsilon s_2}{2\pi i}\int_\alpha^\beta \frac{p(u)}{2u^2}\,\frac{u+s_2}{u-s_2}\,du + \frac{\bar{\varepsilon} s_2}{2\pi i}\int_\alpha^\beta \frac{\bar{p}(u)}{2\bar{u}^2}\,\frac{1+\bar{u}s_2}{1-\bar{u}s_2}\,d\bar{u} +$$

$$+ \gamma\varepsilon - \bar{\gamma}\bar{\varepsilon}s_2^2 + o(\varepsilon).$$

Now

$$\zeta = \psi(s) = h_1 s + h_2 s^2 + \dots$$

maps the arcs Γ, Γ_ε into arcs Γ', Γ'_ε of $|\zeta| < 1$ such that points $s \in \Gamma$, $s + \varepsilon p_\varepsilon(s) \in \Gamma_\varepsilon$ map respectively into $\zeta \in \Gamma'$, $\zeta + \varepsilon q_\varepsilon(\zeta) \in \Gamma'_\varepsilon$ where

$$q_\varepsilon(\zeta) = p_\varepsilon(s)\,\psi'(s) + o(1).$$

Then these points can be brought together by a mapping

$$\zeta_1 = \zeta\left\{1 + \frac{\varepsilon}{2\pi i}\int_a \frac{q(v)}{2v^2}\,\frac{v+\zeta}{v-\zeta}\,dv - \frac{\bar{\varepsilon}}{2\pi i}\int_a \frac{\bar{q}(v)}{2\bar{v}^2}\,\frac{1+\bar{v}\zeta}{1-\bar{v}\zeta}\,dv + O(\varepsilon)\right\}$$

which carries $|\zeta| = 1$ into $|\zeta_1| = 1$. The mapping from s_2 to s to ζ_1 carries the point $s_2 = 0$ into $\zeta_1 = h_1\gamma\varepsilon + o(\varepsilon)$ so let

$$\zeta_2 = \frac{\zeta_1 - \mu\varepsilon}{1 - \bar{\mu}\bar{\varepsilon}\zeta_1}, \quad \mu = h_1\gamma + o(1).$$

Then writing $v = \psi(u)$,

$$\zeta_2 = \psi(s)\left\{1 + \frac{\varepsilon}{2\pi i}\int_\alpha^\beta \frac{p(u)}{2u^2}\left(\frac{u\psi'(u)}{\psi(u)}\right)^2 \frac{\psi(u)+\psi(s)}{\psi(u)-\psi(s)}\,du\right.$$

$$\left. - \frac{\bar\varepsilon}{2\pi i}\int_\alpha^\beta \frac{\bar p(u)}{2\bar u^2}\left(\frac{\bar u\,\bar\psi'(u)}{\bar\psi(u)}\right)^2 \frac{1+\bar\psi(u)\psi(s)}{1-\bar\psi(u)\psi(s)}\,d\bar u\right\} - h_1\gamma\varepsilon + \bar h_1\bar\gamma\bar\varepsilon\psi^2(s) + o(\varepsilon).$$

Combining this with (4) and the mapping $\zeta = \psi(s)$ there results a schlicht mapping from $|s_2| < 1$ onto a domain in the s_2 plane, and, dropping subscripts, this is

$$\psi^*(s) = \psi(s) + \frac{\varepsilon}{2\pi i}\int_\alpha^\beta \frac{p(u)}{2u^2}\left\{\left(\frac{u\psi'(u)}{\psi(u)}\right)^2 \frac{\psi(u)+\psi(s)}{\psi(u)-\psi(s)}\psi(s) - s\,\frac{u+s}{u-s}\psi'(s)\right\}du$$

$$- \frac{\bar\varepsilon}{2\pi i}\int_\alpha^\beta \frac{\bar p(u)}{2\bar u^2}\left\{\left(\frac{\bar u\,\bar\psi'(u)}{\bar\psi(u)}\right)^2 \frac{1+\bar\psi(u)\psi(s)}{1-\bar\psi(u)\psi(s)}\psi(s) - s\,\frac{1+\bar u s}{1-\bar u s}\psi'(s)\right\}d\bar u$$

$$+ \varepsilon\gamma\left(\psi'(s) - h_1\right) + \bar\varepsilon\bar\gamma\left(\bar h_1\psi^2(s) - s^2\psi^1(s)\right) + o(\varepsilon).$$

Writing $\psi^*(s) = h_1^*\,s + \dots$ it follows since $|h_1^*| \geq |h_1|$ that the real part of

$$\frac{2\varepsilon}{2\pi i}\int_\alpha^\beta \frac{p(u)}{2u^2}\left\{\left(\frac{u\psi'(u)}{\psi(u)}\right)^2 - 1\right\}du + 2\,\frac{h_2}{h_1}\,\varepsilon\gamma + o(\varepsilon)$$

must vanish for small ε. Now $p(u)$ is subject to only mild restrictions, so making use of relation (3) it follows that

$$\left(\frac{u\psi'(u)}{\psi(u)}\right)^2 - 1 = \tau\left\{\frac{u+e^{i\phi_1}}{u-e^{i\phi_1}} - \frac{u+e^{-i\phi_1}}{u-e^{-i\phi_1}}\right\}$$

where τ is some constant. Writing s for u, and $\zeta = \psi(s)$, this becomes

$$(4)\qquad \left(\frac{s}{\zeta}\frac{d\zeta}{ds}\right)^2 = \frac{s^2 - 2\lambda s + 1}{(s-e^{i\phi_1})(s-e^{-i\phi_1})}\,.$$

The zeros of the numerator on the right must lie on $|s| = 1$, and indeed on the arc J since $\zeta(s)$ is regular in $|s| < 1$ and on CJ. Thus

$$\int \frac{d\zeta}{\zeta} = \int\left\{\frac{(s-e^{i\phi_2})(s-e^{-i\phi_2})}{(s-e^{i\phi_1})(s-e^{-i\phi_1})}\right\}^{1/2}\frac{ds}{s}$$

where $0 \leqq \phi_2 \leqq \phi_1$ and $\lambda = \cos\phi_2$. If $\phi_2 = \phi_1$ then $\zeta = s$. If

$0 < \phi_2 < \phi_1$ and $s = e^{i\phi}$ then the right side of (4) is real; positive for $0 \leq \phi < \phi_2$, negative for $\phi_2 < \phi < \phi_1$, and positive for $\phi_1 < \phi < \pi$. In this case as s describes $|s| = 1$, the point ζ describes an arc of $|\zeta| = 1$ plus two radial segments seymmetrically placed with respect to the real axis, and an arc of a circle $|\zeta| = r$, where $0 < r < 1$. Since the points $e^{i\phi_1}$, $e^{-i\phi_1}$ of the s plane are not brought together under this mapping it is clear that there is a further mapping of $|\zeta| < 1$ onto a unit circle with a small indentation near the point 1 which would still further decrease the leading coefficient. Thus the case of minimal $|h_1|$ occurs only when $\phi_2 = 0$, in which case the image of $|s| < 1$ under $\zeta = \psi(s)$ is $|\zeta| < 1$ cut by a segment of the positive real axis. Thus $w = f(z)$ maps $|z| < 1$ onto $|w| < 1$ cut by radial segments.

The precise set of points $e^{i\phi}$ of $|z| = 1$ which map into $|w| = 1$ under $w = f(z)$ are those for which every open interval containing $e^{i\phi}$ contains a subset of E of positive measure. This is a perfect subset of E having the same measure as E.

REFERENCES

[1] SCHAEFFER, A. C., SPENCER, D. C., Coefficient regions for schlicht functions, American Mathematical Society Colloquim, Volume 35.

ON DIRICHLET'S PRINCIPLE

MAX SHIFFMAN

1. In his doctoral dissertation, Riemann introduced the Dirichlet principle as a means of constructing analytic functions, or harmonic functions, having prescribed singularities and periods. In particular, the conformal mapping theorem of Riemann would be obtained thereby. The solution was obtained as one which minimizes a specific functional within a class of admissible functions. The existence of a solution to the minimum problem, first proved by Hilbert, could be heuristically seen on physical grounds, since the desired function could be described as one which provides the stationary flow of an ideal incompressible fluid, and the Dirichlet principle related to the minimum principles of mathematical physics.

The real and imaginary parts of an analytic function satisfy the Cauchy-Riemann equations. But it is desirable to be able likewise to discuss solutions of a pair of equations other than the Cauchy-Riemann equations. For example, for a three dimensional flow with rotational symmetry, the velocity potential and the stream function satisfy a system quite similar to the Cauchy-Riemann equations. These are still linear-equations, and it is likewise desirable to be able to discuss functions satisfying a non-linear system of equations. An example of this type of question occurs in the flow of a compressible fluid, and it is this type of question which I shall discuss here.

A particular case of importance is the stationary two dimensional flow of a compressible fluid. The flow is characterized by a velocity potential function $\phi(x,y)$ which makes a certain integral of the following type stationary,

$$\delta \iint G(\phi^2_x + \phi^2_y)dxdy = 0, \tag{1}$$

where $(\phi^2_x + \phi^2_y)^{1/2}$ is the magnitude of the velocity of the fluid at any point and G is a function which depends on the equation of state of the fluid. The resulting partial differential equation is a homogeneous quasi-linear equation which is elliptic when the velocity is smaller than a certain critical value, and hyperbolic when the velocity is larger than this value. There is a stream function $\psi(x,y)$ for the flow determined by

$$(2)\qquad \begin{aligned} G'(\phi^2_x + \phi^2_y) \cdot \phi_x &= \psi_y \\ G'(\phi^2_x + \phi^2_y) \cdot \phi_y &= -\psi_x \end{aligned}$$

The stream function $\psi(x,y)$ satisfies a quasi-linear partial differential equation of the same character as $\phi(x,y)$, which likewise arises from a variational expression

$$(3)\qquad \delta \iint g(\psi^2_x + \psi^2_y)\,dxdy = 0$$

analogous to (1). We thus have a mapping of the (x,y)-plane onto the (ϕ,ψ) plane by a pair of functions which satisfy (2).

To be more definite, consider the flow of an incompressible fluid about a specified object, with a prescribed velocity at infinity. The stream function $\psi(x,y)$ is then constant on the boundary, and the mapping $(x,y) \rightarrow (\phi, \psi)$ can be described as the mapping of the exterior of the given boundary in the (x,y)-plane onto the exterior of a horizontal slit in the (ϕ, ψ)-plane by a pair of functions $\phi(x,y)$, $\psi(x,y)$ satisfying (2).

2. Let B be the region exterior to a given simple closed curve B^*, which is assumed to be a smooth curve, say, with continuous third derivatives. The desired flow in B, in which $\psi_x \rightarrow 0$, $\psi_y \rightarrow q_0$ at infinity, and the desired mapping on the (ϕ, ψ)-plane, will be obtained by a consideration of the variational formulation (3) among func- $\psi(x,y)$ vanishing on the boundary B^*. But the integrand is a regular integrand (furnishes an elliptic partial differential equation as Euler equation) only for $\psi^2_x + \psi^2_y \leqq b^2_0$. We therefore first change the integrand in the following manner. For a fixed positive ϵ, select $b_1^2 = (1 - \epsilon^2)b_0^2$, and define the function $h(\psi^2_x + \psi^2_y)$ to coincide with $g(\psi^2_x + \psi^2_y)$ for $\psi^2_x + \psi^2_y \leqq b_1^2$, is always a convex function of ψ_x, ψ_y, for large $\psi^2_x + \psi^2_y$ is equal to $\alpha + \beta(\psi^2_x + \psi^2_y)$ for suitable constant α, β, and is continuous and has continuous third derivatives. Consider the variational problem for $h(\psi^2_x + \psi^2_y)$ as integrand, which is a regular problem with an elliptic partial differential equation as Euler equation.

The integral is however still divergent since B is an infinite domain, and we consider instead the following minimum problem

$$(4)\qquad \iint_B \Big\{ h(\psi^2_x + \psi^2_y) - h(q_0^2) - 2q_0 h'(q_0^2)(\psi_y - q_0) \Big\}\,dxdy$$
$$= \text{minimum}$$

among all functions $\psi(x,y)$ over B which are continuous, vanish on the boundary B^* of B, and have piecewise continuous first derivative with $\iint_B \{\psi_x^2 + (\psi_y - q_0)^2\}\, dxdy$ convergent. The terms subtracted from $h(\psi_x^2 + \psi_y^2)$ do not affect the Euler equation.

3. The problem (4) is a special case of the variational problem

$$\iint_B \left\{ F(\psi_x,\psi_y) - F(p_0,q_0) - F_p(p_0,q_0)(\psi_x - p_0) - F_q(p_0,q_0)(\psi_y - q_0) \right\} dxdy \quad (5)$$

$$= \text{minimum}$$

among functions ψ vanishing in B^*, where $F(p,q)$ is a convex function of p,q satisfying the inequalities

$$2K(\alpha^2 + \beta^2) \geq F_{pp}(p,q)\alpha^2 + 2F_{pq}(p,q)\alpha\beta + F_{qq}(p,q)\beta^2 \geq 2k(\alpha^2 + \beta^2) \quad (6)$$

for all p,q,α,β where k,K are two positive constants. That this problem has a unique solution $\psi(x,y)$ which has continuous first and second derivatives, and satisfies the Euler equation, can be established on the basis of existing literature.[1] (see bibliography).

As we shall see below, it is of the greatest importance to discuss the behavior of the derivatives ψ_x, ψ_y of the solution on the boundary B^* of B and at infinity. By methods which are based on [6], we prove the following main theorem:

> THEOREM 1. The derivatives ψ_x, ψ_y are continuous in the closure of the domain B, and depend continuously on any parameters appearing in the integrand of (5).

Concerning the behavior at infinity, we know to begin with the convergence of $\iint \{(\psi_x - p_0)^2 + (\psi_y - q_0)^2\}\, dxdy$. We likewise prove that ψ_x and ψ_y are continuous at the point at infinity as well, so that we have $\psi_x \to p_0$, $\psi_y \to q_0$ at infinity. The convergence of the double integral also yields the single-valuedness of the conjugate function $\phi(x,y)$ (corresponding to no circulation).

4. The Euler equation for (4) has therefore been solved; indicate the solution by $\psi(x,y;q_0)$. For $q_0 = 0$, the solution is $\psi(x,y;0) = 0$. This solution varies continuously with q_0, and as a result of the main theorem of § 3, the first derivatives of ψ in the closure of B vary continuously with q_0. Denote the maximum value of $\sqrt{\psi_x^2 + \psi_y^2}$ throughout the domain, which is easily shown to occur on the boundary B^*, by $V(q_0)$. Then $V(q_0)$ varies continuously with q_0, with $V(0) = 0$.

The Euler equation satisfied by $\psi(x,y;q_0)$, corresponding to the integrand of (4), is not necessarily the desired one corresponding to the integrand of (3). But since $h(\ )$ coincides with $g(\ )$ for $\psi^2_x + \psi^2_y \leqq b_1^2$, we see that $\psi(x,y;q_0)$ satisfies the desired equation for the flow of a compressible fluid whenever

$$(7) \qquad V(q_0) \leqq b_1 .$$

This inequality is satisfied for $q_0 = 0$, and by the continuity of $V(q_0)$ in its dependence on q_0 it is therefore satisfied for an interval of q_0. Since b_1 is arbitrarily near b_0, we have obtained the following theorem:

THEOREM 2. There is a critical constant $\overline{q}_0$ such that if $0 < q_0 < \overline{q}_0$, there exists a flow of a compressible fluid about B^*, with $\psi_x = 0$, $\psi_y = q_0$ at infinity and with ϕ single valued. For this flow, denote the maximum of $\sqrt{\psi^2_x + \psi^2_y}$ by $V(q_0)$; then as q_0 varies in $0 < q_0 < \overline{q}_0$, $V(q_0)$ takes all values in $0 < V(q_0) < b_0$, where b_0 is the value of $\sqrt{\psi^2_x + \psi^2_{2y}}$ where the partial differential equation for ψ ceases to be completely elliptic (subsonic flows).[1]

Let us repeat that the boundary B^* was assumed to be smooth.

5. In terms of the mapping on the (ϕ,ψ)-plane, we see that there is a mapping of the region B onto the region exterior to a horizontal slit in the (ϕ,ψ)-plane, by a pair of functions satisfy (2), with the point at infinity mapped onto the point at infinity, and for a range of distortions at infinity corresponding to $0 < q_0 < \overline{q}_0$.

For the case of the flow of a compressible fluid, it is also desirable to consider the case of circulation in addition. In this case, the subtractive terms in (4) or in (5) will be chosen differently. In (5), let $S(x,y)$ be a solution of

$$(8) \qquad F_{pp}(p_0,q_0)S_{xx} + 2F_{pq}(p_0,q_0)S_{xy} + F_{qq}(p_0,q_0)S_{yy} = 0$$

which is of the form $S = p_0x + q_0y + \sigma \log R$ where R^2 is the square of distance from a point interior to B^* in the plane obtained from (x,y) by the linear transformation which transforms the elliptic equation (8) to the Laplace equation. Then in place of (5) consider

[1] An analogous theorem for minimal surfaces has been obtained by L. Bers, Trans. Amer. Math. Soc., vol. 70 (1951), pp. 465-491, using the continuity method of Schauder-Leray.

$$\int_B\int \left\{ F(\psi_x,\psi_y) - F(S_x,S_y) - A(x,y)(\psi_x - S_x) - B(x,y)(\psi_y - S_y) \right\} dxdy \tag{9}$$
$$= \text{minimum}$$

where

$$\left.\begin{aligned} A(x,y) &= F_p(p_0,q_0) + F_{pp}(p_0,q_0)(S_x-p_0) + F_{pq}(p_0,q_0)(S_y-q_0) \\ B(x,y) &= F_q(p_0,q_0) + F_{qp}(p_0,q_0)(S_x-p_0) + F_{qq}(p_0,q_0)(S_y-q_0) \end{aligned}\right. . \tag{10}$$

Then the subtracted terms do not affect the Euler equation by virtue of

$$\frac{\partial A}{\partial x} + \frac{\partial B}{\partial y} = 0;$$

and also

$$\int_B\int [A - F_p(S_x,S_y)]^2\, dxdy \quad , \quad \int_B\int [B - F_q(S_x,S_y)]^2\, dxdy < \infty$$

and these guarantee a solution to (9), and all of the preceding considerations. A similar theorem to §4 is obtained, with now $\overline{q}_0$ depending on the circulation constant σ as well.

Similar considerations apply to any number of boundaries, or other kinds of flows or partial differential equations

BIBLIOGRAPHY

[1] HAAR, A., Math. Ann., vol. 97 (1927), pp. 124-158.
[2] TONELLI, L., Ann. Scuola norm. super. Pisa, II s. 2 (1933), pp. 89-130.
[3] HOPF, E., Math. Zeits., vol. 30 (1929), pp. 404-413.
[4] MORREY, C. B., Trans., Amer. Math. Soc., vol. 43, (1938), pp. 126-166.
[5] MORREY, C. B., Univ. of Calif., Publications in math., new ser., vol. 1 (1943), pp. 1-130.
[6] SHIFFMAN, M., Ann. of Math., vol. 48 (1947), pp. 274-284.

A PROBLEM CONCERNING THE CONTINUATION OF RIEMANN SURFACES

MAURICE HEINS

1. A Riemann surface F is said to admit a Riemann surface G as a <u>continuation</u> [1,3] provided that there exists a (1,1) directly conformal map of F onto a region of G. In this paper we consider the following problem:

What can be said about a Riemann surface which admits all compact Riemann surfaces of given positive genus as continuations?

Our basic result is the following theorem:

THEOREM 1. A necessary and sufficient condition that a Riemann surface F admit continuation to all compact Riemann surfaces of a given positive genus is that F be conformally equivalent to a bounded plane region.

The sufficiency part of the theorem is of course trivial. To establish the necessity we proceed in two steps. First, we show that, if F satisfies the given hypothesis, then F admits continuation to all compact Riemann surfaces of genus one. We thereupon treat the problem for the case of genus one with the aid of uniformization methods

From an intuitive point of view the theorem is not surprising when one takes into account the possibilities of degeneration that can occur. The basic problem is to study the influence which such degenerations may exert. In this connection we point out that in the proofs the full force of the hypothesis that F admits continuation to all compact Riemann surfaces of assigned positive genus is not used. For the case of genus one we shall replace the hypothesis in question by a weakest possible one. The appropriate formulation for the case of genus greater than one deserves further study.

2. In this and the following section we dispose of several preliminary considerations which will be wanted for the main argument. The first is the following lemma.

LEMMA 1. A planar subregion of a compact Riemann surface of positive genus is conformally equivalent to a bounded plane region.

PROOF. Let p denote the genus of the given surface. We assert that the given subregion is contained in a second planar region of the surface whose connectivity does not exceed $2p$. The frontier of the latter region must contain a component which is a continuum. Hence the latter region possesses a Green's function and must be conformally equivalent to a bounded plane region. The lemma then follows.

To establish the assertion, we proceed as follows. Let R denote the given region, F the surface, $\{R_n\}_0^\infty$ and exhaustion of R where frR_n consists of the union of a finite number, m_n, of disjoint analytic closed Jordan curves and $\overline{R}_n \subset R_{n+1}$, $n = 0,1,\ldots$. Further let the components of the complement of R_n with respect to F be $\Phi_{1,n},\ldots,$ $\Phi_{r_n,n}$ Then with X denoting the Euler characteristic, we have

$$\sum_{1}^{r_n} X(\Phi_{1,n}) + m_n - 2 = 2p - 2. \tag{2.1}$$

Letting μ_n denote the number of $\Phi_{1,n}$ with $X(\Phi_{1,n}) = -1$, we have

$$\mu_n \geq m_n - 2p. \tag{2.2}$$

Such $\Phi_{1,n}$ are simply-connected. On adjoining them to R_n we obtain a planar subregion R_n^* of connectivity $\leq 2p$. We note that $\{R_n^*\}$ is increasing and that the connectivity of R_n^* is a non-decreasing function of n and hence is ultimately constant. Let $R^* = \lim R_n^*$. Then R^* has the desired properties.

3. Let F denote a given Riemann surface and let G denote a given compact Riemann surface of genus one. We may endow G in a natural manner with a metric $d[q_1,q_2]$ with the aid of a uniformization mapping g of the finite plane onto ("the universal covering surface of") G. By definition $d[q_1,q_2]$ is the minimum of the euclidean distances between antecedents of q_1 and antecedents of q_2 with respect to g. With the aid of this metric we may assign an obvious meaning to such statements as: "a sequence of maps f_n of F into G tends uniformly in F to a map f of F into G." If the f_n are directly conformal and tend uniformly to f, then f is directly conformal or constant.

Let $\mathcal{F}$ denote the family of directly conformal maps f of F into G which are at most m-valent $(m \geq 1)$. It is assumed that $\mathcal{F}$ is not empty. The family $\mathcal{F}$ is normal in the sense that given a sequence $\{f_n\}_0^\infty$ of members of $\mathcal{F}$ there exists a subsequence $\{f_{k(n)}\}_0^\infty$ which tends uniformly in F to a map of F into G. To see this, let s_0 denote a given point of F and let R denote a simply-connected region of F containing s_0. Given $f \in \mathcal{F}$, there exists a unique analytic function φ with domain R specified by the requirements:

(1) $\varphi(s_0)$ lies in a fixed fundamental parallelogram of g.

(2) $g[\varphi(s)] = f(s), \quad s \in R$.

We note that the area of the Riemannian image of R with respect to φ is dominated by m times the area of the fundamental parallelogram. Hence the family of such φ is normal in R and we conclude that the f restricted to R constitute a normal family. It is thereupon readily concluded that $\mathcal{F}$ is normal in F.

A further property of the family $\mathcal{F}$ that will be useful is that, if $f_n(\in \mathcal{F}) \rightarrow f(\neq \text{const.})$ uniformly in F, then $f \in \mathcal{F}$.

4. Suppose now that F is a Riemann surface which admits continuation to all compact Riemann surfaces of given genus p (> 1). We show that F admits continuation to all compact Riemann surfaces of genus one. To that end let G denote a given compact Riemann surface of genus one, let $q_0 \in G$ and let $\eta\,(> 0)$ be chosen so small that the set of q satisfying $d'q_0,q] \leq \eta$ is homeomorphic to a closed circular disc. We construct a Riemann surface H_q of genus p and a $(1,p)$ directly conformal map of H_q onto G in the following manner. (It is to be emphasized that the specific nature of the construction will play a role in the proof that follows.)

For q satisfying $0 < d[q_0,q] \leq \eta$ we take p copies of G slit along the rectilinear (in the sense of the metric) segment $\overline{q_0q}$, say $G_1,\dots,G_p$ and distinguish in the same manner for all the copies "positive" and "negative" sides for the slit $\overline{q_0q}$ and thereupon join the G_k along the slits in the customary fashion, the positive edge of G_k being identified with the negative edge of G_{k+1} $(k \bmod p)$. In this way an orientable surface H_q of genus p and a $(1,p)$ interior map of this surface onto G have been constructed. The mapping so given induces a conformal structure on its domain and in the sense of this conformal structure the mapping is directly conformal. We remark that the mapping is ramified precisely over q_0 and q for each of these points the ramification index is $p-1$. Further as a consequence of the specific mode of construction, the mapping is univalent on each component of the antecedent of G less $\overline{q_0q}$.

The composition of the constructed mapping with a $(1,1)$ directly conformal mapping of F into H_q is a directly conformal mapping of F into G which is at most p-valent. For each q considered, let f_q denote such a composed map of F into G.

We may put aside the case where F is planar by virtue of the lemma of Section 2. If F is not planar, there exists a relatively compact subregion of F, say F^*, which is not planar. For every nonseparating retrosection γ of G less the segment $\overline{q_0q}$ we have

$$f_q(F^*) \cap \gamma \neq \emptyset \; . \tag{4.1}$$

Otherwise F^* would be conformally equivalent to a subregion of H_q less the p retrosections antecedent to γ and would hence be planar.

There exist a sequence of q, say $\{q_n\}_1^\infty$ and an associated sequence of maps $\{f_{q_n}\}_1^\infty$ such that $q_n \to q_o$ and $\{f_{q_n}\}_1^\infty$ converges uniformly in F. Let f denote the limit map. We assert that f is univalent. We note first that for F^* as above and γ_k (k=1,2) two disjoint non-separating retrosections of G less $\overline{q_o q}$ (γ_k independent of q), (4.1) is in effect with γ_k taking over the role of γ. Hence $f(\overline{F^*}) \cap \gamma_k \neq \emptyset$ (k=1,2) and it follows that f is not identically constant.

If f were not univalent, there would exist distinct points $s_1, s_2 \in F$ such that $f(s_1) = f(s_2) (\neq q_o)$. Let F^* now denote a relatively compact subregion of F containing s_1, s_2. For n sufficiently large there would exist $s_1^{(n)}$, $s_2^{(n)}$ in assigned disjoint neighborhoods of s_1, s_2 respectively such that $f_{q_n}(s_k^{(n)}) = f(s_1) = f(s_2) (k=1,2)$. It follows from the above construction of H_q and the associated mapping that f_{q_n} would attain on F^* a point of the segment $\overline{q_o q_n}$. Hence f would attain q_o at a point of $\overline{F^*}$, say s_o. Consider a neighborhood N of s_o such that f is univalent at each point of $N - \{s_o\}$. For n sufficiently large, q_o and q_n would be attained by f_{q_n} in N and f_{q_n} would be of multiplicity p at the antecedents of q_o and q_n. We infer that f would have multiplicity $2p-1$ ($>p$) at s_o. This is clearly impossible since f is at most p-valent. Hence f is univalent and our assertion is thereby established. We note that f cannot attain q_o.

Our problem is reduced to the case where F admits continuation to all compact Riemann surfaces of genus one. It is to be observed that it is sufficient to show that F is planar by virtue of the lemma of Section 2. As we have remarked, we treat this phase of the problem by uniformization methods. One would be tempted to continue in the spirit of the present section by using two-sheeted elliptic covering surfaces of the extended plane, but it is not clear how one would achieve a desirable element of compactness for the family of mappings considered.

5. Let F now denote a Riemann surface admitting a proper continuation to a compact Riemann surface of genus one and let φ denote a fixed uniformization mapping of $|z| < 1$ onto F. For each compact Riemann surface T of genus one to which F admits continuation, let f denote a univalent directly conformal mapping of F into T and let ψ denote a uniformization mapping of the finite plane onto T which is so normalized that $\psi(0) = f[\varphi(0)]$. There then exists* a unique function g which is analytic in $|z| < 1$ and satisfies the conditions

$$f \circ \varphi = \psi \circ g, \quad g(0) = 0. \tag{5.1}$$

* For other applications of this method, cf. [2].

Since the function g is locally simple, we may further assume that ψ has been so chosen that $g'(0) = 1$. Let $\mathfrak{F}_F$ denote the Fuchsian (or Fuchsoid) group of linear fractional transformations which leave φ invariant and let Ω denote the group of the periods of ψ. We denote a generic member of $\mathfrak{F}_F$ by σ and a generic member of Ω by ω. It is immediate that

$$g \circ \sigma = g + \omega_\sigma, \quad \sigma \in \mathfrak{F}_F, \tag{5.2}$$

and that, if $g(z_1) \equiv g(z_2) (\mathrm{mod}\, \Omega)$, then $z_1 \equiv z_2 \ (\mathrm{mod}\ \mathfrak{F}_F)$ (i.e., there exists $\sigma \in \mathfrak{F}_F$ such that $z_2 = \sigma z_1$). This latter observation follows from the univalence of f. From (5.2) we see that g induces a homomorphic mapping of $\mathfrak{F}_F$ into Ω. Since $g'(0) = 1$ and g is univalent in the interior of a fixed fundamental polygon of $\mathfrak{F}_F$ containing the origin in its interior, it follows from the Koebe $\frac{1}{4}$-theorem that the moduli of the non-zero periods of Ω have a positive lower bound independent of Ω.

On the other hand, if g is analytic in $|z| < 1$ and Ω is an additive group of complex numbers such that (1) g induces a homomorphic mapping of $\mathfrak{F}_F$ into Ω in the sense of (5.2) and (2) $g(z_1) \equiv g(z_2) (\mathrm{mod}\, \Omega)$ implies $z_1 \equiv z_2$ (mod $\mathfrak{F}_F$) then the moduli of the non-zero members of Ω have a positive lower bound (possibly $+\infty$) and F admits continuation to a Riemann surface T (finite plane, punctured plane, torus) such that Ω is the group of periods of a suitably chosen uniformization mapping of the finite plane onto T.

We consider the family Γ of g defined by (5.1) and subject to the normalization $g'(0) = 1$, taking into account all T and f. We note that as a consequence of the normalization of the g at the origin and the uniform local univalence* of the g, the family Γ is normal in $|z| < 1$.

Suppose that g_0 is the limit of a sequence $\{g_n\}_1^\infty$ of functions of Γ which converges uniformly in $|z| < 1$. We assert that g_0 satisfies the following conditions:

(1) g_0 induces a homomorphic mapping of $\mathfrak{F}_F$ <u>onto</u> an additive group Ω_0 of complex numbers in the sense indicated above,

(2) $g_0(z_1) \equiv g_0(z_2)$ (mod Ω_0) implies $z_1 \equiv z_2$ (mod $\mathfrak{F}_F$).

The first condition follows from the fact that $g_n \circ \sigma - g_n$ is identically constant for each $n, \sigma (\in \mathfrak{F}_F)$. The second condition may be verified as follows. If $g_0(z_1) = g_0(z_2) + \omega, \omega \in \Omega_0$, there exists $\sigma_\omega \in \mathfrak{F}_F$ such that $\omega_n = g_n \circ \sigma_\omega - g_n \rightarrow \omega$. Since g_0 is univalent in some

---

* i.e. There exists a neighborhood of each point of $|z| < 1$ in which each $g \in \Gamma$ is univalent.

neighborhood of z_2, there exists a sequence $\{z_2^{(n)}\}$ such that $z_2^{(n)} \to z_2$ and

$$g_n(z_1) = g_n(z_2^{(n)}) + \omega_n$$

for n sufficiently large. Hence since $z_2^{(n)} \equiv z_1 \pmod{\mathcal{B}_F}$ for n sufficiently large, we have $z_2 \equiv z_1 \pmod{\mathcal{B}_F}$.

6. As is well-known, the conformal type of a torus T is specified by the principal module of its associated period groups, that is, the module $\mu(T)$ satisfying the conditions

$$-\frac{1}{2} \leq R\mu < \frac{1}{2}; \quad \mathcal{I}\mu > 0; \quad |\mu| \geq 1 \quad \text{for } -\frac{1}{2} \leq R\mu \leq 0;$$
$$|\mu| > 1 \quad \text{for } 0 < R\mu < \frac{1}{2}.$$

We are now in a position to prove the following theorem.

THEOREM 2. If F is a Riemann surface of genus one, the set of the principal modules of the tori to which F admits continuation is bounded.

Before turning to the proof, we remark that it is an immediate consequence of this theorem and the lemma of Section 2 that, if the set of principal modules of the tori to which a Riemann surface F admits continuation is unbounded (in particular, if F admits continuation to all tori), then F is conformally equivalent to a bounded plane region.

Suppose the theorem not true. There would exist a sequence of tori $\{T_n\}_1^\infty$ with $\lim \mu(T_n) = \infty$ and each T_n a continuation of F. For each n, let g_n denote the analytic function in $|z| < 1$ associated with a univalent directly conformal mapping of F into T_n according to the specifications of Section 5. We may as well assume that g_n converges uniformly in $|z| < 1$ and we denote the limit function by g.

Let (α_n, β_n) denote a pair of generators of the period group of the uniformization of T_n associated with g_n, so chosen that $\beta_n = \mu(T_n)\alpha_n$.* We note that since F is not planar, the homomorphic image of $\mathcal{B}_F$ induced by g_n is not cyclic. Let (γ_n, δ_n) denote a pair of generators of this image group, so chosen that $\gamma_n^{-1}\delta_n$ is its principal module. We have

* We recall that if (α, β) is such a pair of generators with $\alpha^{-1}\beta$ the principal module, then α is a non-zero period of minimum modulus.

$$\begin{aligned} \gamma_n &= A_n \alpha_n + B_n \beta_n \\ \delta_n &= C_n \alpha_n + D_n \beta_n \end{aligned} \tag{6.1}$$

where A_n, B_n, C_n, D_n are real integers.

Since F is not planar, it follows that $|\gamma_n|$ is bounded. On noting that $|\alpha_n|$ has a positive lower bound independent of n, we infer on dividing the first equation of (6.1) by α_n and then taking imaginary parts that $B_n = 0$ for n sufficiently large. For such n, $D_n \neq 0$ since F is not planar. It follows on dividing the second equation of (6.1) by α_n and thereupon taking imaginary parts that $\lim \alpha_n^{-1} \delta_n = \infty$ and hence $\lim \gamma_n^{-1} \delta_n = \infty$.

On the other hand, let (γ, δ) denote a pair of generators for the homomorphic image of $\mathcal{B}_F$ induced by g, with $\gamma^{-1}\delta$ the principal module of this image group (not cyclic). For n sufficiently large, the homomorphic image of $\mathcal{B}_F$ induced by g_n contains periods Γ_n, Δ_n which are arbitrarily close to γ and δ respectively. Hence for n large

$$\left|\frac{\delta_n}{\gamma_n}\right| \leq \max\left(\left|\frac{\Gamma_n}{\gamma_n}\right|, \quad \left|\frac{\Delta_n}{\gamma_n}\right|\right) = O(1)$$

since $|\gamma_n|$ has a positive lower bound. This constitutes a contradiction and the theorem follows.

We remark that in the converse direction, if the set of the principal modules of the tori to which F admits continuation is bounded (it being assumed that such tori exist), then F is of genus one.

7. We turn now to an inverse problem. Suppose that $\mathcal{T}$ is a family of tori T whose principal modules constitute a bounded set. Does there exist a Riemann surface of genus one which admits all $T \in \mathcal{T}$ as continuations? We shall see that the answer is affirmative. The proof is based upon the classical result which states that T admits a two-sheeted mapping onto the extended plane whose ramification points are precisely $0, 1, \infty, \lambda(\mu(T))$ where λ denotes the customarily so designated modular function. Now the set of $\mu(T), T \in \mathcal{T}$, is contained in a set of the following form

$$|Rz| \leq \tfrac{1}{2}, \ |z| \geq 1, \ 0 < \mathcal{J}z \leq A \quad (A > 1).$$

On this set λ is univalent. The complement of its λ-image with respect to the extended plane is a simply-connected region containing $0, 1, \infty$. Let F denote a Riemann surface which admits a two-sheeted mapping onto this region which is ramified over $0, 1, \infty$ but over no other points of the region. Then F is clearly of genus one and admits continuation to all $T \in \mathcal{T}$.

8. An alternative proof of the theorem of Section 6 may be given as follows. It uses more precise information than the above proof but by way of compensation it reveals that the homomorphic mappings induced by the g are actually onto mappings.

If F is a non-compact Riemann surface of genus one, then there exist on F two non-separating retrosections τ_1 and τ_2 having precisely one point in common at which they "cross" one another. We take this point as $\varphi(0)$. It may be shown that the fundamental group of T based at $f[\varphi(0)]$ is generated by the path classes containing the f-images of paths τ_1' and τ_2' with initial and terminal point at $\varphi(0)$ which are obtained from τ_1 and τ_2 respectively in an obvious way. Let $z_k(t)$, $0 \leq t \leq 1$ $(k=1,2)$ denote a continuous representative of the antecedent of τ_k' with respect to φ with $z_k(0) = 0$. Then $z_k(1) = \sigma_k 0$ for unique $\sigma_k \in \mathcal{J}_F$. It may be concluded from the above property of $f(\tau_k')$ $(k=1,2)$ that $\omega_{\sigma_1}, \omega_{\sigma_2}$ generate Ω.

On taking into account all admitted T and associated g, we see that the sets

$$\omega_{\sigma_k} \qquad (k=1,2)$$

are bounded by virtue of the normality of $|g|$. Let (α,β) denote a generating pair of the Ω associated with a given g such that $\alpha^{-1}\beta$ is the principal module of Ω. Recall that $|\alpha|$ has a positive lower bound independent of Ω. Since

$$|\beta| \leq \max\,(|\omega_{\sigma_1}|,\ |\omega_{\sigma_2}|),$$

we infer that the set of $\mu(T)$ is bounded and the theorem follows.

BIBLIOGRAPHY

[1] BOCHNER, S., Fortsetzung Riemannscher Flächen. Math. Ann. Bd. 98 (1927) pp. 406-421.

[2] HEINS, M., On the continuation of a Riemann surface. Annals of Math. Vol. 43 No. 2, pp. 280-297.

[3] RADO, T., Ueber eine nicht fortsetzbare Riemannsche Mannigfaltigkeit. Math. Zeit. Bd. 20 (1924). pp. 1-6.

CONSTRUCTION OF FUNCTIONS WITH PRESCRIBED PROPERTIES ON RIEMANN SURFACES

LEO SARIO

In this survey, two methods are discussed, a linear method and an extremal method, for constructing functions with preassigned properties on arbitrary Riemann surfaces. Also, a summary is made of the present knowledge on the classification problem of Riemann surfaces. In the bibliography, a list of papers published thus far (1948-1951) is given.

§1. HISTORICAL NOTE

The first rigorous proof of Riemann's fundamental mapping theorem was given by H. A. Schwarz, using a method known as the alternating method. Since this method is frequently also connected with the name of C. Neumann (e.g. [20]), it is perhaps of interest to recall the exact historical circumstances.

Riemann had based the proofs in his Inaugural Dissertation, 1851, on Dirichlet's principle. For nearly two decades his argument was generally accepted. In particular, the first edition of Neumann's book on Abelian integrals, 1865, was entirely based on Dirichlet's principle. Then Weierstrass, in July, 1870, announced his famous counter-example, showing that no rigorous proof had been given for Dirichlet's principle and that Riemann's results, therefore, were not established.

Schwarz, who also had expressed doubts about Dirichlet's principle, described his alternating method in April, 1870, in a lecture before the Naturforschende Gesellschaft in Zürich. His detailed paper on the method was introduced by Weierstrass in October, 1870, to the Königliche Akademie der Wissenschaften in Berlin. At the end of the paper, Schwarz stated that he had, therewith, given the first complete proof of Riemann's results.

So far the priority question is clear. But now comes the confusing point. Neumann published (1884) the second edition of his book on Abelian integrals. In it he abandoned the Dirichlet principle and presented all his proofs by "gürtelförmige Verschmelzung". However, in the whole book there is no mention of Schwarz or his method. About this, we quote Schwarz in his Werke II, pp. 361-362: "Dieselbe Grenze hat Herr Carl Neumann in der zweiten Auflage seiner 'Vorlesungen über Riemann's Theorie der Abelschen Integrale' auf andere, ihm eigenthümliche Weise hergeleitet und diese Grenzbestimmung zur Grundlage für einen Grenzübergang gemacht, für welchen er die Benennung 'gürtelförmige Verschmelzung' vorgeschagen hat. Dieser Grenzübergang stimmt mit demjenigen Übergange genau überein, welchen der Verfasser in der im Jahre 1870 von Herrn Weierstrass der Königlichen Akademie der Wissenschaften zu Berlin vorgelegten Abhandlung in zwei Fällen angewendet hat. In beiden Fällen entspricht dem Gürtel des Herrn Carl Neumann der von den beiden Linien L_1 und L_2 begrenzte, mit T^* bezeichnete zweifach zusammenhängende Bereich."

The lack of reference to Schwarz was corrected by Neumann four years later in Ber. Sächs. Ges. Wiss. Leipzig 40 (1888), p. 122: "Bei meinem dortigen Beweise der Riemannschen Existenztheoreme spielt diejenige combinatorische Methode, welche ich daselbst als die der gürtelförmigen Verschmelzung bezeichnet habe, eine hervorragende Rolle. -- Um so mehr sehe ich mich verpflichtet, zu bemerken, dass diese Methode (wie ich vor Kurzem durch eine Mitteilung von H. A. Schwarz erfahren habe) schon vor mir von H. A. Schwarz gefunden und publiziert worden ist, in seiner Abhandlung von 1870, in den Berichten der Berliner Akademie."

Thus the priority question was completely settled by the two rivals themselves. The alternating method is thus due to Schwarz, for annular intersections as well as for overlapping circles.

§2. CONSTRUCTION METHODS ON ARBITRARY RIEMANN SURFACES

The alternating method was used by Schwarz and Neumann for constructing functions on closed Riemann surfaces. That the method applies also to parabolic surfaces, is obvious (no. 3 in [40]), since the boundary in this case has no effect on the behaviour of alternating functions. But there are essentially two points in Schwarz's reasoning which prevent using his method for arbitrary Riemann surfaces. First, if the two overlapping domains have large (ideal) boundaries, the harmonic functions on them are no longer determined by their values on the relative boundaries. Secondly, Schwarz's convergence proof fails because Poisson's integral has no meaning on arbitrary domains. By certain changes these obstacles can be overcome [38, 51]. But a shorter and more powerful method results, if, starting from Schwarz's idea of alternation, all inessential requirements on the approximating functions are dropped and a large class of different limit functions is constructed simultaneously. For a formal simplification of this procedure certain linear operators are used. This linear method leads further to an extremal method which has independent applications to construction problems.

We will discuss these methods in search for a unified treatment of the following general

> PROBLEM. Given an arbitrary Riemann surface, construct a function with a prescribed minimal property m in a prescribed class F of functions.

The minimizing function will be termed the principal function $p(m,F)$ in the class F with respect to m.

We shall first select m so as to characterize the boundary behaviour of functions. As results, we will obtain relations between boundary properties and existence properties on a given surface R. Then we select m to be a combination of a power series coefficient and a boundary integral. This will give us relations between extremal coefficients and complementary areas and lead to a new approach to the classification problem of Riemann surfaces.

§3. FUNCTIONS WITH PRESCRIBED BOUNDARY BEHAVIOUR

We begin by giving a precise meaning to our first problem. Let R be an arbitrary Riemann surface. It may be of finite or infinite genus; it may be given as a covering surface of the complex plane, abstractly by a set of parameter discs, or as a group of linear transformations. In any case we can use as its topological model a 2-dimensional surface imbedded in a 3-dimensional space.

We must first of all give a precise meaning to the expression "a prescribed boundary behaviour". To this end we describe the conventional notion "a prescribed singularity" in such terms that it can be applied to a boundary neighborhood.

By "a prescribed singularity" we mean this: On a neighborhood S_1 of a point P of R there is given a harmonic function s_1 with a singularity at P. The function p to be constructed is said to have the singularity s_1 in S_1, if the difference $p-s_1$ is harmonic in S_1. This property can be expressed as follows.

Let v be a single-valued harmonic function in some neighborhood of the boundary α of S_1. Denote by Lv the harmonic function in S_1 with the boundary values v on α. Here L is an operator which clearly satisfies the following simple conditions:

$$Lv = v \text{ on } \alpha, \text{ by definition,} \tag{1}$$

$$\int_\alpha d\overline{Lv} = 0, \tag{2}$$

$$\min_\alpha v \leq Lv \leq \max_\alpha v, \tag{3}$$

$$L(c_1v_1 + c_2v_2) = c_1Lv_1 + c_2Lv_2. \tag{4}$$

Here, as everywhere in this paper, the barred letters stand for conjugate harmonic functions.

We call an operator L satisfying these conditions a normal linear operator. The property that $p-s_1$ is harmonic in S_1 can now be expressed by the relation

$$p-s_1 = L(p-s_1) \text{ in } S_1;$$

that is, the difference $p-s_1$ in S_1 is furnished by a normal linear operator L from its boundary values on α.

It is this property that we now transfer to a boundary neighborhood. For such a neighborhood we can select the complementary region S_2 of S_1. Let s_2 be a harmonic function in S_2 with an arbitrary growth towards the boundary of R. We say that the function p to be constructed has the prescribed boundary behaviour s_2 in S_2 if the difference $p-s_2$ in S_2 satisfies the condition

$$p-s_2 = L(p-s_2).$$

By (3), this means, in particular, that the difference $p-s_2$ in S_2 attains its maximum and minimum on the relative boundary α; that is, the function p actually imitates the growth of s_2, regardless of how strongly the latter increases.

By this definition, we have attained a complete uniformity for the notion of a prescribed singularity and prescribed boundary behaviour. Thus, we can ignore the boundary and put the problem as follows.

On a Riemann surface R let α be an analytic Jordan curve and $S = S_1 + S_2$ its complement. In S let L be a normal linear operator and s a harmonic function with arbitrary singularities (s consists of two parts, in S_1 and S_2). The problem is to construct on R a harmonic function p with the behaviour s; that is, such that

$$p-s = L(p-s) \text{ in } S.$$

§4. LINEAR OPERATOR METHOD

To start with, we note the following necessary condition. Suppose we have constructed the desired function p. Naturally,

$$\int_{\alpha^+ + \alpha^-} d\bar{s} = 0$$

where α^+ and α^- are the different edges of α, traced in opposite directions. Since, by (2), also

$$\int_{\alpha^+ + \alpha^-} d\overline{L(p-s)} = 0,$$

we have the necessary condition

$$\int_{\alpha^+ + \alpha^-} d\bar{s} = 0.$$

We will see that this simple condition is not only necessary but also sufficient. It fully characterizes the solvability of our construction problem.

The underlying idea of our method will be as follows. The two branches of s in general differ from each other on α. We will successively add small correction functions to these branches until they are finally harmonic continuations of each other, thus forming the desired function p on the whole surface.

To this end, let S' be a doubly connected domain, bounded by analytic Jordan curves $\beta_1 \subset S_1$ and $\beta_2 \subset S_2$ such that α divides S' into two conformally equivalent annuli.

We start with the function

$$s_0 = s - Ls$$

in S. Let s' be the harmonic function in S' which coincides with s_0 on $\beta = \beta_1 + \beta_2$. Using the values of s' on α, form the function Ls' on S. As a function of the values of s_0 on β, it is clearly a linear operator Ks_0. Denote by K^i the ith iterate of K, and consider the sum

$$s_n = \sum_{i=1}^{n} K^i s_0. \tag{6}$$

The functions $K^i s_0$ are successive corrections, as we mentioned above, of the original function s_0. The convergence of the sequence follows now from the inequality

$$\max | K^{i+1} s_0 | \leq q \max | K^i s_0 | , \tag{7}$$

where the constant $0 < q < 1$ is independent of i. This inequality can be easily derived from the condition (5) (see [47]). The limit

$$p = s_0 + \lim s_n \tag{8}$$

is the desired function on R. In fact, since, by construction, $s_n = Ls_n$, we obtain the required relation $p-s = L(p-s)$ by letting $n \to \infty$.

Thus we have the following result [47]:

THEOREM 1. Let R be an arbitrary Riemann surface and α an analytic Jordan curve on R. In $S = R - \alpha$, let s be a single-valued function, both branches of which are harmonic on α. The condition

$$\int_{\alpha^+ + \alpha^-} d\bar{s} = 0, \tag{9}$$

is necessary and sufficient for the existence of a single-valued real function p on R, harmonic on α and such that

$$p-s = L(p-s) \tag{10}$$

on S. The function p is constant if and only if $s \equiv Ls$.

We note that we have obtained a method which is valid for arbitrary Riemann surfaces; the closed surfaces are contained in our treatment as a special case.

Also, the same method is applicable to different construction problems. In fact, there are in general several operators L which satisfy the conditions (1) - (4). This gives us a double freedom. Firstly, we can select the singular function s. And secondly we are free to select the law L by which the function p is related to s.

For every subdomain D (with analytic relative boundary α) of an arbitrary Riemann surface R, there is, in particular, a linear operator L_λ which associates with any single-valued harmonic function v on α, a harmonic function u_λ in D such that the expression

$$m_\lambda(u) = \int_\beta u\, d\bar{u} + \lambda \int_\alpha u\, d\bar{u} \tag{11}$$

is minimized by $u \equiv u_\lambda$ [50]. Here β is the ideal boundary of D and λ a real parameter $(-1 \leq \lambda \leq 1)$; u ranges in the class of single-valued harmonic functions in D with $u = v$ on α, $\int_\alpha d\bar{u} = 0$. For $\lambda = -1$, the operator L_λ minimizes the Dirichlet integral of u. Theorem 1 yields necessary and sufficient conditions, in terms of this operator, for the existence of certain harmonic or analytic functions on R [47].

§5. EXTREMAL METHOD. THE P-SPAN AND THE Q-SPAN

We now turn to the extremal method mentioned in §2. Consider the class $\{P\}$ of analytic functions on an arbitrary open Riemann surface R with the development

$$P = p + i\bar{p} = \frac{1}{z} + \sum_1^\infty a_\nu z^\nu \tag{12}$$

in a parameter disc $K: |z| \leq 1$ and with a single-valued p on R; write $\alpha = \operatorname{Re}(a_1)$. The functions P are supposed to have no singularities other than at $z = 0$. As an expression m to be minimized, we will now take a combination of α and the integral $\int p\, d\bar{p}$ extended along the (ideal) boundary β of R.

Let $\{F\}$ be the class of analytic functions on R with a single-valued real part, and with the development

$$F = f + i\bar{f} = \sum_1^\infty b_\nu z^\nu, \quad \operatorname{Re}(b_1) = 1 \tag{13}$$

in K. We have [48]:

THEOREM 2. There is in the class $\{P\}$ a uniquely determined principal function

$$P_\lambda = p_\lambda + i\bar{p}_\lambda = \frac{1}{z} + \sum_1^\infty a_{\lambda\nu} z^\nu \tag{14}$$

which minimizes the expression

$$m_\lambda (p) = 2\pi\lambda\alpha + \int_\beta p d\bar{p}. \tag{15}$$

The function P_1 minimizes, P_{-1} maximizes α among functions $|P|$ with $\int_\beta p\, d\bar{p} \leq 0$.

The P-span

$$\sigma_P = \alpha_{-1} - \alpha_1, \tag{16}$$

where $\alpha_\lambda = \text{Re}\,(a_{\lambda 1})$, gives the extrema

$$\sigma_P = -\min \frac{2}{\pi} \int_\beta p d\bar{p} = -\frac{2}{\pi} \int_\beta p_o d\bar{p}_o \tag{17}$$

and

$$\frac{1}{\sigma_P} = \min \frac{1}{2\pi} D(f) = \frac{1}{2\pi} D(f_o), \tag{18}$$

with

$$p_o = \tfrac{1}{2}(p_{-1} + p_1)$$

and

$$f_o = \frac{1}{\sigma_P} (p_{-1} - p_1).$$

PROOF. Suppose first that R has an analytic boundary. Let p_1 and p_{-1} be the functions p determined by

$$p_1 = \text{const. on } \beta \tag{19}$$

$$\frac{\partial p_{-1}}{\partial n} = 0 \text{ on } \beta. \tag{20}$$

The existence of these functions is guaranteed by §4. Let

$$p_\lambda = \frac{1+\lambda}{2} p_1 + \frac{1-\lambda}{2} p_{-1}. \tag{21}$$

A straightforward calculation (where the integrals along β are first transferred by Green's formula to integrals along $|z| = 1$, and these determined by the residue theorem) gives

$$m_\lambda (p) = \tfrac{1}{2}\pi[(1+\lambda)^2\alpha_1 - (1-\lambda)^2\alpha_{-1}] + D(p - p_\lambda). \tag{22}$$

This proves the theorem for a finite R.

If R is arbitrary, a limiting procedure [48] shows that the important formula (22) remains valid, and the theorem follows.

Theorem 2 holds also for the subclass $|Q|$ of $|P|$ defined by the restriction that $\int d\bar{q} = 0$ along every cycle dividing R into disjoint parts. The P-span is to be replaced by the Q-span $\sigma_Q = \alpha_{-1} - \alpha_1$, where now the extrema of α are taken with respect to $\{Q\}$. For planar surfaces the functions Q are single-valued analytic functions and the Q-span coincides with Schiffer's span. The well-known results of de Possel, Grunsky and Schiffer concerning extrema of α and $\int q\, d\bar{q}$ are re-established in a unified manner. From the boundedness of $p_{-1} - p_1$ and $q_{-1} - q_1$ (for planar surfaces $Q_{-1} - Q_1$) the classification results known at present on the basis of existence of regular functions on arbitrary Riemann surfaces immediately follow [48].

§6. THE CAPACITY OF THE BOUNDARY AND OF A BOUNDARY COMPONENT

Our extremal method can also be used to define the capacity of the boundary and of a boundary component of an abstract Riemann surface. No use will be made of the exhaustion of the surface.

Consider, on an arbitrary open Riemann surface R, the class $\{S\}$ of analytic functions with the development

$$S = s + i\bar{s} = \log z + \sum_{1}^{\infty} a_\nu z^\nu \tag{23}$$

in K: $|z| \leq 1$ and with a single-valued real part s on R.

DEFINITION 1. The capacity c_β of the boundary β of an open Riemann surface is

$$c_\beta = e^{-k_\beta} \tag{24}$$

with

$$k_\beta = \min \frac{1}{2\pi} \int_\beta s\, d\bar{s} \tag{25}$$

where s ranges in the above class $\{s\}$.

The proof of the existence of a function s_β which minimizes $\int s\, d\bar{s}$ is given in [49] where relations to other existence problems are also derived. Now let γ be a boundary component of an open Riemann surface R. This notion, intuitively clear, can be defined in exact terms [49]. Consider the subclass $\{T\}$ of $\{S\}$,

$$T = t + i\bar{t} = \log z + \sum_{1}^{\infty} b_\nu z^\nu, \tag{26}$$

determined by the restriction $\int d\bar{t} = 2\pi$ along every cycle separating the (fixed) boundary component γ from $z = 0$ and $\int d\bar{t} = 0$ along other cycles. We introduce:

DEFINITION 2. The capacity c_γ of a boundary component γ of an open Riemann surface R is

$$c_\gamma = e^{-k_\gamma} \tag{27}$$

with

$$k_{\gamma} = \min \frac{1}{2\pi} \int_{\beta} t \, d\bar{t}. \tag{28}$$

The capacity of γ has significance in the classification problem of Riemann surfaces (§8).

§7. HISTORICAL NOTE ON THE CLASSIFICATION PROBLEM

By his mapping theorem, based upon the existence or non-existence of the Green's function, Riemann introduced the classification problem of Riemann surfaces and gave the fundamental division into two types which are called parabolic and hyperbolic. In the last section of his Inaugural Dissertation, Riemann specifically stated that his reasoning was by no means restricted to the simply connected case but was valid for arbitrary surfaces.

The notion (not the name) "harmonic measure" was introduced by H. A. Schwarz in 1890 on p. 360 of his Gesammelte Werke, vol. II. There $\chi/2\pi$ is the harmonic measure, used for giving to the Poisson integral the elegant form $\frac{1}{2\pi} \int f(\vartheta) d\chi$. The current notation $\omega(z, \gamma, D)$ for the harmonic measure was introduced by Beurling in his Dissertation (Upsala 1933) on p. 26. The name, the "harmonic measure", is due to Nevanlinna (Stockholm, 1934) who applied this notion in the theory of Riemann surfaces. The class of "nullbounded" surfaces coincides with Riemann's class of surfaces without Green's functions, as was shown by Myrberg (Acta Math., 1933) for surfaces of finite genus and in [39, 51] for the general case. The theory of Abelian integrals was generalized to parabolic surfaces by Nevanlinna (Ann. Acad. Sci. Fenn. A. I. 1 (1941)).

The early works on these subjects were apparently guided by the erroneous impression that the "nullboundary" was decisive in related existence and uniqueness problems (Ann. Acad. Sci. Fenn. A. I. 1 (1941), p. 31; ibid. 56 (1949), p. 8). In the Congress of Scandinavian Mathematicians in Copenhagen, 1946, Nevanlinna reported on a work by the present writer, contained in two manuscripts (April and August, 1946). In the first it was noted that the uniqueness of the Schottky-uniformization of closed surfaces had been demonstrated in the course of Courant's proof of the Kreisnormierungsprinzip (Hurwitz-Courant, Funktionentheorie, p. 516) and that in this proof the type of the Schottky covering surface did not enter. This seemed to indicate that, contrary to the opinion current at that time, in certain closely related problems, the "nullboundary" did <u>not</u> constitute the dividing line. Repeated efforts to find the "nullboundary" hiding in Courant's proof failed. (By 1951, no rigorous determination of the type of the Schottky-surface has been published).

Courant's reasoning was implicitly based on the non-existence of analytic functions with a finite Dirichlet-integral on the Schottky-surface. This fact led the author to the following general classification problem, suggested, together with some results, in the second manuscript:

> PROBLEM. To find the classes of open surfaces which share with closed surfaces the property of not admitting certain regular single-valued functions.

To indicate this property the writer introduced the terms surfaces of closed character or surfaces with removable boundary (cf. [56]).

In [35], the surfaces with removable boundary with respect to analytic functions with a finite Dirichlet integral were studied in detail. In particular, a criterion was derived, in terms of the divergence of the modular product of an exhaustion. The existence of hyperbolic surfaces with removable boundary was established. It was proved that the boundary of every Abelian covering surface of a closed Riemann surface is removable. The problem concerning the connection with bounded analytic functions was stated on p. 76, and the corresponding classification problem for harmonic functions on p. 7. In [36], an extension of the problem was given for functions in a boundary neighborhood and for functions with a bounded mean value; a unified notation was introduced.

Since the Copenhagen Congress, the general classification question was also attacked by Ahlfors who had earlier investigated the Painlevé problem [Duke Math. J. 14 (1947)]. In a related paper [1] he solved extremal problems on Riemann surfaces and extended Nevanlinna's theory of Abelian integrals to surfaces which do not allow harmonic functions with a finite boundary integral. Later he succeeded in showing that this class is effectively larger than the class of parabolic surfaces. For plane regions the complete classification was obtained jointly by Ahlfors and Beurling in [3].

During the past three years, the classification question has been vigorously investigated by a growing number of authors. A list of the papers published thus far is given in the bibliography. In the sequel we shall assemble the principal results.

8. RIEMANN SURFACES WITH REMOVABLE BOUNDARY

We use the following abbreviations for certain properties of functions. First, periodicity properties:

H harmonic single-valued non-constant,

K harmonic single-valued non-constant, the imaginary part having no periods along dividing cycles,

A analytic single-valued non-constant,

W meromorphic non-constant,

M analytic non-constant with a single-valued modulus.

Then, more special properties:

S univalent (schlicht) single-valued non-constant,

P belonging to class $\{P\}$ (§5),

Q belonging to class $\{Q\}$ (§5).

Finally, boundedness properties:

B bounded,

C positive,

D with a finite Dirichlet integral,

E omitting a set of values of positive area,

N with a negative boundary integral $\int p\, d\bar{p}$.

By a combination of letters we denote classes of functions with the corresponding combination of properties [36].

By O_F, we will denote the class of Riemann surfaces R with F-<u>removable boundary</u> [35], i.e. with no functions belonging to a given class F on R. We will compare these classes mutually and with the following classes defined by conformally invariant measures:

S_P the P-span vanishes,

S_Q the Q-span vanishes,

C_β the capacity of β vanishes.

C_γ the capacity of each boundary component γ vanishes.

The following relations have been found thus far:

$$O_A \subset C_\beta = O_{MB} = O_{MD} \subseteq O_{HC} \subseteq O_{HB} \subseteq S_P = O_{PN}$$
$$= O_{HBD} = O_{HD} \subset S_Q = O_{QN} = O_{KBD} = O_{KD} \subseteq O_{AD}. \tag{29}$$

Furthermore, it is known that,

$$O_{AB} = O_{WE} \subset O_{AD}, \tag{30}$$

but the relation of O_{AB} to O_{HD} and to O_{KD} is unknown for surfaces with arbitrary genus. For planar surfaces, the additional relations

$$C_\beta = O_{HC} = O_{HB} = O_{HD} \subset O_{AB} \subset S_D = O_{ABD}$$
$$= O_{AD} = O_{SE} \subset C_\gamma = O_{SB} = O_{SD} \tag{31}$$

hold.

The relations concerning classes of functions H, A and S (schlicht) were found by Ahlfors and Beurling [3], Behnke and Stein [5], Florack [7], Lehto [13], Lokki [14], Parreau [29], Royden [33] and Virtanen [54]. The relations concerning M, S_P, S_Q, P, Q, K, C_γ were established in [48, 49], where a unified proof of all the above relations is also given.

As to the connection of the above classes of planar surfaces with the classes M_λ of surfaces with a vanishing λ-dimensional Hausdorff measure of the boundary (in all its realizations as a point set on the complex plane), the following relations, found by Ahlfors [Duke Math. J. 14, p. 2], Myrberg [17], Strebel [53] and Lehto [13], are known:

$$C_\beta \subset M_\varepsilon \subset M_1 \subseteq O_{AB} \subset M_{1+\varepsilon} \subset M_{2-\varepsilon} \subset O_{AD} = M_2. \tag{32}$$

We note that all the classes considered above are in one of the arbitrarily narrow intervals (M_0, M_ε), $(M_1, M_{1+\varepsilon})$, $(M_{2-\varepsilon}, M_2)$.

Heins [9] found recently a subclassification of parabolic surfaces. Bader and Parreau [4], and Nevanlinna [22] have given relations concerning functions in subdomains with non-compact relative boundaries. Parreau [26] has investigated the existence of harmonic functions with bounded mean values. The Japanese school (Kuroda, Mori, Nagai, Nagura, Noshiro, Ohtsuka, Ozawa, Tsuji) has made contributions on related problems, especially existence criteria, and simplified earlier proofs.

BIBLIOGRAPHY

In this bibliography we have tried to include the literature published thus far (1948-1951) on the general classification problem. A list of earlier papers on the classical type problem was published by Le-Van in Comm. Math. Helv. 20 (1947), p. 286.

[1] AHLFORS, LARS, Open Riemann surfaces and extremal problems on compact subregions, Comm. Math. Helv. 24 (1950), pp. 100-134.

[2] -"-, Remarks on the classification of open Riemann surfaces, Ann. Acad. Sci. Fenn. A. I. 87 (1951), pp. 1-8.

[3] AHLFORS, LARS, and BEURLING, ARNE, Conformal invariants and function-theoretic null-sets, Acta Math. 83 (1950), pp. 101-129.

[4] BADER, ROGER and PARREAU, MICHEL, Domaines non compact et classification des surfaces de Riemann, C. R. Acad. Sci. Paris 232 (1951), pp. 138-139.

[5] BEHNKE, HEINRICH and STEIN, KARL, Entwicklung analytischer Funktionen auf Riemannschen Flächen, Math. Ann. 120 (1947-49), pp. 430-461.

[6] CHRISTOFFEL, MARTIN, Ueber eine Klasse offener Ueberlagerungsflächen, Zürich, Leeman (1950), pp. 1-42.

[7] FLORACK, HERTA, Reguläre und meromorphe Funktionen auf nichtgeschlossenen Riemannschen Flächen, Schriftenreihe Math. Inst. Univ. Münster, Heft. 1 (1948), pp. 1-33.

[8] HEINS, MAURICE, The conformal mapping of simply connected Riemann surfaces, Annals of Math. 50 (1949), pp. 686-690.

[9] -"-, Riemann surfaces of infinite genus, Annals of Math. (to appear).

[10] KURODA, TADASHI, On the type of an open Riemann surface, Proc. Jap. Acad. 27 (1951), pp. 57-60.

[11] -"-, Some remarks on an open Riemann surface, Tohoku Math. J. (2) 3 (1951), pp. 182-186.

[12] -"-, Notes on an open Riemann surface, Kodai Math. Sem. Rep. 3-4 (1951), pp. 61-63.

[13] LEHTO, OLLI, On the existence of analytic functions with a finite Dirichlet integral, Ann. Acad. Sci. Fenn. A. I. 67 (1949), pp. 1-6.

[14] LOKKI, OLLI, Beiträge zur Theorie der analytischen und harmonischen Funktionen mit endlichem Dirichlet Integral, Ann. Acad. Sci. Fenn. A. I. 92 (1951), pp. 1-10.

[15] MORI, AKIRA, On Riemann surfaces on which no bounded harmonic function exists, J. Math. Soc. Jap. (to appear).

[16] -"-, On the existence of harmonic functions on a Riemann surface, J. Fac. Sci. Univ. Tokyo, Section I, vol. VI (1951), pp. 247-257.

[17] MYRBERG, P. J., Ueber die analytische Fortsetzung von beschränkten Funktionen, Ann. Acad. Sci. Fenn. A. I. 58 (1949), pp. 1-7.

[18] NAGAI, YASUTAKA, On the behaviour of the boundary of Riemann surfaces, Proc. Jap. Acad. 26 (1950).

[19] NAGURA, SHOHEI, Kernel functions on Riemann Surfaces, Kodai Math. Sem. Rep. (1951), pp. 73-76.

[20] NEVANLINNA, ROLF, Ueber die Neumannsche Methode zur Konstruktion von Abelschen Integralen, Comm. Math. Helv. 22 (1949), pp. 302-316.

[21] -"-, Sur l'existence de certaines classes de différentielles analytiques, C. R. Acad. Sci. Paris 228 (1949), pp. 2002-2004.

[22] -"-, Ueber die Existenz von beschränkten Potentialfunktionen auf Flächen von unendlichem Geschlecht, Math. Z. 52 (1950), pp. 559-604.

[23] NOSHIRO, KIYOSHI, Open Riemann surfaces with null-boundary, Nagoya Math. J. 3 (1951), pp. 73-79.

[24] OTHSUKA, MAKOTO, Dirichlet problems on Riemann surfaces and conformal mapping, Nagoya Math. J. 3 (1951), pp. 91-137.

[25] OZAWA, MITSURU, On classification of the function-theoretic null-sets on Riemann surfaces of infinite genus, Kodai Math. Sem. Rep. (1951), pp. 43-44.

[26] PARREAU, MICHEL, Sur les moyennes des fonctions harmoniques et la classification des surfaces de Riemann, C. R. Acad. Paris 230 (1950), pp. 42-44.

[27] -"-, Comportement à la frontière de la fonction de Green d'une surface de Riemann, C. R. Acad. Sci. Paris 230 (1950), pp. 709-711.

[28] -"-, La théorie du potentiel sur les surfaces de Riemann à frontière positive, C. R. Acad. Sci. Paris 230 (1950), pp. 914-916.

[29] -"-, Sur certaines classes de fonctions analytiques uniformes sur les surfaces de Riemann, C. R. Acad. Sci. Paris 231 (1950), pp. 751-753.

[30] PFLUGER, ALBERT, La croissance des fonctions analytiques et uniformes sur une surface de Riemann ouverte, C. R. Acad. Sci. Paris 229 (1949), p. 505-507.

[31] -"-, Ueber das Anwachsen eindeutiger analytischer Funktionen auf offenen Riemannschen Flächen, Ann. Acad. Sci. Fenn. A. I. 64 (1949), pp. 1-18.

[32] -"-, Sur l'existence de fonctions non constantes, analytiques, uniformes et bornées sur une surface de Riemann ouverte, C. R. Acad. Sci. Paris, 230 (1950), pp. 166-168.

[33] ROYDEN, H. L., Some remarks on open Riemann surfaces, Ann. Acad. Sci. Fenn. A. I. 85 (1951), pp. 1-8.

[34] -"-, Harmonic functions on an open Riemann surface, Trans. Amer. Math. Soc. (to appear).

[35] SARIO, LEO, Ueber Riemannsche Flächen mit hebbarem Rand, Ann. Acad. Sci. Fenn. A. I. 50 (1948), pp. 1-79.

[36] -"-, Sur la classification des surfaces de Riemann, Congr. Math. Scand. Trondheim 1949, pp. 229-238.

[37] -"-, Sur le problème de type des surfaces de Riemann, C. R. Acad. Sci. Paris 229 (1949), pp. 1109-1111.

[38] -"-, Existence des fonctions d'allure donnée sur une surface de Riemann arbitraire, C. R. Acad. Sci. Paris 229 (1949), pp. 1293-1295.

[39] -"-, Quelques propriétés à la frontière se rattachant a la classification des surfaces de Riemann, C. R. Acad. Sci. Paris 230 (1950), pp. 42-44.

[40] -"-, Existence des intégrales abéliennes sur les surfaces de Riemann arbitraires, C. R. Acad. Sci. Paris 230 (1950), pp. 168-170.

[41] -"-, Questions d'existence au voisinage de la frontière d'une surface de Riemann, C. R. Acad. Sci. Paris 230 (1950), pp. 269-271.

[42] -"-, On open Riemann surfaces, Int. Congr. Math. Cambridge 1950, pp. 398-399.

[43] -"-, Alternating method for analytic functions. Bull. Amer. Math. Soc. 57 (1951), pp. 275-276.

[44] -"-, Linear operators on Riemann surfaces, Bull. Amer. Math. Soc. 57 (1951), p. 276.

[45] -"-, Principal functions on Riemann surfaces, Bull. Amer. Math. Soc. 57 (1951), pp. 475-476.

[46] -"-, Existence of functions as a boundary property, Bull. Amer. Math. Soc. 58 (1952), p. 45.

[47] -"-, A linear operator method on arbitrary Riemann surfaces, Trans. Amer. Math. Soc., 72 (1952), pp. 281-295.

[48] -"-, An extremal method on arbitrary Riemann surfaces, Trans. Amer. Math. Soc. 72 (1952), pp. 459-470.

[49] -"-, Capacity of the boundary and of a boundary component, Bull. Amer. Math. Soc. 58 (1952), (to appear in extenso, Annals of Math.).

[50] -"-, Minimizing operators on subregions, Proc. Amer. Math. Soc. (to appear in 1953).

[51] -"-, Alternating method on arbitrary Riemann surfaces, Pacific J. Math. (to appear in 1953).

[52] -"-, Modular criteria on Riemann surfaces, Duke Math. J. (to appear in 1953).

[53] STREBEL, KURT, Eine Bemerkung zur Hebbarkeit des Randes einer Riemannschen Fläche, Comm. Math. Helv. 23 (1949), pp. 350-352.

[54] VIRTANEN, K. I., Ueber die Existenz von beschränkten harmonischen Funktionen auf offenen Riemannschen Flächen, Ann. Acad. Sci. Fenn. A. I. 75 (1950), pp. 1-7.

[55] TSUJI, MASATSUGU, Some theorems on open Riemann surfaces, Nagoya Math. J. 3 (1951), pp. 141-145.

[56] Zentralblatt fur Mathematik 32 (1950), p. 474.

METRIC RIEMANN SURFACES[1]

E. CALABI

1. INTRODUCTION

The complex analytic structure of a Riemann surface is introduced in F. Klein's book [4] by considering the isothermal parameters in a differentiable, orientable surface in Euclidean 3-space. This method of course can be applied more generally to any abstract 2-dimensional, orientable Riemannian manifold. Admittedly this purely auxiliary metric structure on the Riemann surface is somewhat artificial, as evidenced by the fact that, beyond the proof of existence of local isothermal parameters, the function theory on the Riemann surface on one hand and the Riemannian geometry of the surface on the other seem to come essentially to a parting of the ways (it is significant that both theories should bear the same man's name). However, the rapid and fruitful development of the theory of Kähler manifolds in the last twenty years demonstrates how much can be obtained from the natural interconnection between the two structures.

The present note is a specialization to the 2-dimensional case, with additional results that are characteristic of it, of the author's doctoral thesis [3], which deals precisely with problems of isometric, complex analytic imbedding of complex manifolds. The purpose is to characterize all metrics that can be induced on a Riemann surface $\mathfrak{M}$ by a regular analytic map into the unitary N-dimensional space C^N $(N \leq \infty)$. If $x_1, \ldots, x_\sigma, \ldots$ $(1 \leq \sigma \leq N)$ are unitary coordinates in C^N such that the line element is $ds^2 = \Sigma_{\sigma=1}^N |d\, x_\sigma|^2$, and if z is a local uniformizer of $\mathfrak{M}$, this is equivalent to considering sets of holomorphic functions $x_\sigma = f_\sigma(z)$, whose first derivatives vanish nowhere simultaneously. In case $N = \infty$ we must also assume that the map of $\mathfrak{M}$ into C^N is continuous with respect to the Hilbert space topology of C^∞, for it has not been possible to prove it from complex analyticity alone.[2] The continuity of the map implies that the real-valued function $\Phi(z,\overline{z}) = \Sigma_{\sigma=1}^N |f(z)|^2$ is bounded in the neighborhood of any point; from this one proves that $\Phi(z,\overline{z})$ is real analytic using Lebesgue's theorem on dominated convergence and Weierstrass's theorem on uniform limits of holomorphic functions. Thus one sees that the first fundamental form induced on $\mathfrak{M}$, given by

$$ds^2 = \Sigma_{\sigma=1}^N |dx_\sigma|^2 = g(z,\overline{z})\ |dz|^2 \tag{1}$$

where

$$g(z,\bar{z}) = \Sigma_{\sigma=1}^{N} |f'_{\sigma}(z)|^2 = \frac{\partial^2 \Phi(z,\bar{z})}{\partial z \, \partial \bar{z}}$$

is a real analytic, positive density of weight 1. In this paper we not only characterize all metrics $g|dz|^2$ that can be thus obtained, but indicate also, in the course of the proofs, how one can obtain all possible generating functions $f_{\sigma}(z)$, if the necessary conditions are satisfied.

One can similarly characterize the metrics induced on $\mathfrak{M}$ by mapping it into the Fubini-Study spaces (cfr. [2], [3]); the latter spaces are especially interesting in the elliptic case, for they are the only ones of this type into which one can imbed a compact Riemann surface. However, this part will be omitted, as the statements and proofs do not present any new essential difficulties.

2. UNITARY GEOMETRY OF COMPLEX CURVES

We shall understand by a unitary coordinate transformation in C^N a substitution of the type $x'_{\sigma} = \Sigma_{\tau=1}^{N} a_{\sigma\tau} x_{\tau} + b_{\sigma}$, where $a_{\sigma\tau}$ is a unitary $N \times N$ matrix; by a unitary motion we mean a point transformation in C^N that carries a coordinate frame into one similarly related to the stationary frame. The fundamental form, invariant under the group of motions is

$$ds^2 = \Sigma_{\sigma=1}^{N} |dx_{\sigma}|^2.$$

If a region $\mathcal{R}$ of a Riemann surface $\mathfrak{M}$ admitting a local uniformizer z is mapped regularly and analytically in C^N by

$$x_{\sigma} = f_{\sigma}(z) = \Sigma_{\mu=0} a_{\sigma,\mu} z^{\mu}, \tag{2}$$

and p is any point in $\mathcal{R}$, one can replace $z(q)$ in (2) by $z(q) - z(p)$; if $\mathcal{R}$ does not lie in any proper linear surface of C^N, there is a unique unitary transformation of coordinates into the <u>canonical coordinates</u> of C^N with respect to p and the uniformizer z; they are characterized as follows: $x_{\sigma}(z(p)) = 0$, hence $a_{\sigma,0} = 0$ for all σ and if $a_{\sigma,\ell_{\sigma}}$ is the first non-vanishing coefficient in the power series of (2), then $a_{\sigma,\ell_{\sigma}}$ is real and positive and $\ell_{\sigma} \geq \ell_{\sigma-1} + 1 \geq \sigma$. Since the mapping is regular, we have $a_{1,1} = \sqrt{g(0,0)} > 0$, i.e., $\ell_1 = 1$. The integers ℓ_{σ} are independent of the local uniformizer and so are the constants $a_{\sigma,\ell_{\sigma}} (a_{1,1})^{-\ell_{\sigma}}$. The canonical coordinates with respect to a given point p are uniquely determined regardless of the local uniformizer up to a substitution from the toroidal group,

$$x'_{\sigma} = e^{i\theta_{\sigma}} x_{\sigma}.$$

The map of $\mathcal{R}$ into the right coset space of unitary transformations modulo the toroidal subgroup is analytic. The point p is called <u>ordinary</u> if each

$\ell_\sigma = \sigma$; otherwise it is called a point of hyper-osculation; the latter points form at most a countable set in $\mathcal{R}$. For a fixed uniformizer z in $\mathcal{R}$ and variable base point p ranging over the ordinary points alone, the canonical coordinates in C^N with respect to p and z depend on p continuously.

We now define formally the higher curvatures $\rho_\sigma (1 \leq \sigma < N)$ of $\mathcal{R}$ at any point p, in complete analogy with the case of real curves in Euclidean space, in terms of the coefficients of (2), where x_σ are the canonical coordinates with respect to p, by

$$\rho_\sigma(p) = K \frac{a_{\sigma+1, \ell_{\sigma+1}}}{a_{1,1} a_{\sigma, \ell_\sigma}},$$

where

$$K = (1 + \frac{1}{\ell_\sigma}) \prod_{\mu=1}^{\sigma-1} (1 + \frac{1}{\ell_\sigma - \ell_\mu}) .$$

The point p is ordinary if and only if $\rho_\sigma(p) \neq 0$ $(\sigma = 1,2,\ldots,N-1)$.

The curvatures $\rho_\sigma(p)$ are continuous, real scalars, i.e., they are independent of the local uniformizers; their squares are indeed real analytic functions. The totality of the functions ρ_σ together with the metric $g(z,\bar{z})$ determine the map of $\mathcal{R}$ into C^N up to a unitary motion. The essential property that shows the contrast between real curve theory and complex curve theory, however, is that, while in the case of real curves the curvatures ρ_σ are arbitrary differentiable functions, here they are all dependent on the metric g, as we shall prove.

3. CONDITIONS FOR ISOMETRIC IMBEDDING

We have already indicated that the metric induced on $\mathcal{M}$ by a regular analytic imbedding in C^N is necessarily real analytic. There are only two other conditions imposed besides the one of analyticity; the first, a purely local one, is developed in the following theorems; the second one is a global one that insures the single-valuedness of the mapping functions.

THEOREM 1. Let $\mathcal{M}$ be a metric Riemann surface. For any point p_0 of $\mathcal{M}$ there exists a neighborhood $\mathcal{R}$ of p_0 that can be isometrically and complex analytically imbedded in the Hilbert space C^∞ if and only if the metric is, first of all, real analytic and, secondly, if z is a local uniformizer on $\mathcal{R}$ at p, and (1) is the metric, then the real functions $G_\nu(z,\bar{z})$ $(\nu = 0,1,2,\ldots)$ are all non-negative in $\mathcal{R}$, where

$$G_0(z,\bar z) = 1; \quad G_1(z,\bar z) = g(z,\bar z)$$

(3)
$$G_{\nu+1}(z,\bar z) = \frac{\left(G_\nu(z,\bar z)\right)^2}{G_{\nu-1}(z,\bar z)} \quad \frac{\partial^2}{\partial z\,\partial \bar z} \log G_\nu(z,\bar z) \qquad (\nu \geq 1).$$

PROOF. We assume first that a neighborhood $\mathcal{R}$ of p_0 defined by $|z| < \rho$ $(z(p_0) = 0)$, can be isometrically imbedded in C^N by (2). Since the metric $g(z,\bar z)$ is real analytic, we can write

$$g(z,\bar z) = \Sigma_{\mu,\nu=1}^{\infty} A_{\mu\nu} z^{\mu-1} \bar z^{\nu-1} ,$$

where

(4)
$$A_{\mu\nu} = \mu\nu \Sigma_{\sigma=1}^{\infty} a_{\sigma,\mu} \bar a_{\sigma,\nu} = \frac{1}{(\mu-1)!(\nu-1)!} \frac{\partial^{\mu+\nu-2} g(z,\bar z)}{\partial z^{\mu-1} \partial \bar z^{\nu-1}} .$$

The second member of (4) clearly indicates that the infinite matrix $A_{\mu\nu}$ is hermitian and positive semi-definite; on denoting by $g_{\mu,\nu}$ the derivative

$$\frac{\partial^{\mu+\nu} g(z,\bar z)}{\partial z^{\mu} \partial \bar z^{\nu}} ,$$

this implies at least that at the point p_0 the minor determinants of rank μ

$$\begin{vmatrix} g(p_0) & \cdots & g_{0,\mu-1}(p_0) \\ g_{1,0}(p_0) & \cdots & g_{1,\mu-1}(p_0) \\ \cdots & \cdots & \cdots \\ g_{\mu-1,0}(p_0) & \cdots & g_{\mu-1,\mu-1}(p_0) \end{vmatrix} = \prod_{\nu=1}^{\mu-1} (\nu!)^2 \cdot \begin{vmatrix} A_{11} & \cdots & A_{1\mu} \\ \vdots & & \vdots \\ A_{\mu 1} & \cdots & A_{\mu\mu} \end{vmatrix}$$

must be non-negative for each value of μ. An easy computation shows that these minors are precisely the values of the functions G_μ of (3) at p_0. Since the same argument is valid at every point $p \in \mathcal{R}$ and since the functions $G_\mu(z,\bar z)$ depend on the uniformizer as real densities of weight $\mu(\mu+1)$, we have proved both the necessity and invariance of the condition.

Conversely suppose that each $G_\mu(z,\bar z)$ is analytic and non-negative. Then the set of points p where each G_μ is either positive or the function itself identically zero, being the complement of a set of first Baire category, is everywhere dense in $\mathcal{R}$ and the domains of convergence of the power series of each G_μ around each such point p cover $\mathcal{R}$. Let p_0 be one such point, z a uniformizer vanishing at p_0, and

$$g(z,\bar z) = \Sigma_{\mu,\nu=1}^{\infty} A_{\mu\nu} z^{\mu-1} \bar z^{\nu-1};$$

if none of the functions G_μ vanish, then $A_{\mu\nu}$ is a positive definite matrix; if G_N is the last non-vanishing member of the sequence, then the matrix $A_{\mu\nu}$ is positive semi-definite of rank N and the minor $|A_{\mu\nu}|_{\mu,\nu\leq N}$ is positive. In either case the matrix can be written uniquely in the form

$$\Sigma_{\sigma=1}^{N}\ \mu\nu a_{\sigma,\mu}\bar{a}_{\sigma,\nu} \quad ,$$

where $a_{\sigma,\mu} = 0$ for $\mu < \sigma$ and $a_{\sigma,\sigma}$ is real and positive. Then we have

$$g(z,\bar{z}) = \Sigma_{\mu,\nu=1}^{\infty}\ (\Sigma_{\sigma=1}^{N}\ a_{\sigma,\mu}\bar{a}_{\sigma,\nu})\ z^{\mu-1}\ \bar{z}^{\nu-1} = \Sigma_{\sigma=1}^{N}\left|\Sigma_{\mu=\sigma}^{\infty}\mu a_{\sigma,\mu}\ z^{\mu-1}\right|^2 ,$$

where the change of order in the summation is permissible by the absolute convergence of power series. It is clear that, if we set

$$f_\sigma(z) = \Sigma_{\mu=\sigma}^{\infty} a_{\sigma,\mu}\ z^{\mu},$$

then

$$g(z,\bar{z}) = \Sigma_{\sigma=1}^{N}\ |f_\sigma'(z)|^2 \quad ,$$

so that the functions $x_\sigma = f_\sigma(z)$ are analytic in the same domain as the domain of convergence of the power series of $g(z,\bar{z})$ and represent canonical coordinate of the points of that domain with respect to p_0 under an isometric, complex analytic imbedding of the domain into C^N, thus completing the proof of Theorem 1.

We now prove that the isometric imbedding of Riemann surfaces in C^N is essentially unique.

THEOREM 2. Let $\mathcal{R}$ be any domain in a metric Riemann surface $\mathcal{M}$ that can be isometrically and analytically mapped into C^N so that the image does not lie in any proper linear subspace of C^N. Then N is uniquely determined and the map is unique up to a unitary motion in C^N.

PROOF. If

$$g(z,\bar{z}) = \Sigma_{\mu,\nu=1}^{\infty}\ A_{\mu\nu}\ z^{\mu-1}\ \bar{z}^{\nu-1},$$

we let

$$\Phi(z,\bar{z}) = \Sigma_{\mu,\nu=1}^{\infty}\ \frac{1}{\mu\nu}\ A_{\mu\nu}\ z^{\mu}\ \bar{z}^{\nu} \quad .$$

Then $\Phi(z,\bar{z})$ is a real valued, analytic scalar satisfying

$$\frac{\partial^2 \Phi(z,\bar{z})}{\partial z \partial \bar{z}} = g(z,\bar{z}) .$$

This last equation determines Φ uniquely modulo harmonic functions. If p, q are any two points in the domain U of convergence of the power series of Φ, we let

$$D(p,q) = \Phi\left(z(p), \overline{z(p)}\right) + \Phi\left(z(q), \overline{z(q)}\right) - \Phi\left(z(p), \overline{z(q)}\right) - \Phi\left(z(q), \overline{z(p)}\right) .$$

This functional element generates a symmetric function in a subregion of $\mathfrak{M} \times \mathfrak{M}$, called the <u>diastasis</u> of $\mathfrak{M}$; it is uniquely determined by the metric and represents, for any fixed q, a normalization of the functions $\Phi\left(z(p), \overline{z(p)}\right)$, since

$$\frac{\partial^2 D(p,q)}{\partial z \partial \bar{z}}(p) = g\left(z(p), \overline{z(p)}\right) .$$

If $x_\sigma(p) = f_\sigma\left(z(p)\right)$ are functions representing an isometric map of $\mathcal{R}$ into C^N and $y_\tau(p) = h_\tau\left(z(p)\right)$ another set of functions mapping $\mathcal{R}$ similarly into $C^{N'}$, then

$$D(p,q) = \Sigma_{\sigma=1}^{N} \left| f_\sigma\left(z(q)\right) - f_\sigma\left(z(p)\right) \right|^2 = \Sigma_{\tau=1}^{N'} \left| h_\tau\left(z(q)\right) - h_\tau\left(z(p)\right) \right|^2 .$$

It follows that finite Euclidean distances between the images of p and q in C^N and $C^{N'}$ respectively are identical; the conclusion of the theorem then follows from well known facts about Euclidean geometry. The numbers N and N' are also proved equal by the fact that both are equal to the rank of the matrix $A_{\mu,\nu}$ at least when the origin is an ordinary point.

We call a metric Riemann surface <u>resolvable of rank N</u> at a point p if the metric is analytic at p and if some neighborhood of p can be isometrically mapped into C^N but not in $C^{N'}$ for any $N' < N$. We next prove that if a metric Riemann surface is resolvable of rank N at any one point and if the metric is analytic, then first of all the metric is resolvable of rank N everywhere and from that we prove that the universal covering space $\tilde{\mathfrak{M}}$ of $\mathfrak{M}$ can be globally mapped into C^N.

THEOREM 3. If a Riemann surface with an analytic metric is resolvable of rank N at any one point p, then it is resolvable of rank N everywhere.

PROOF. It is clear that the set $\mathcal{R}$ of points where the metric is resolvable of rank N is open; hence all we have to prove is that it is also closed.

Let p_0 be the limit point of $\mathcal{R}$ and z a uniformizer at p_0

such that $z(p_0) = 0$. Consider the power series expansion of the diastasis $D(p_0,p) = \Sigma_{\mu,\nu=1}^{\infty} B_{\mu\nu} (z(p))^{\mu} (\overline{z(p)})^{\nu}$, which we assume to converge for p in some domain U. Then we can expand the matrix $B_{\mu\nu}$ of coefficients in the form

$$B_{\mu\nu} = \Sigma_{\sigma=1}^{\infty} b_{\sigma,\mu} \overline{b}_{\sigma,\nu} - \Sigma_{\sigma=1}^{\infty} c_{\sigma,\mu} \overline{c}_{\sigma,\nu}$$

in some way, so that the functions

$$h_{\sigma}(z) = \Sigma_{\mu=1}^{\infty} b_{\sigma,\mu} z^{\mu}$$

and

$$h_{-\sigma}(z) = \Sigma_{\mu=1}^{\infty} c_{\sigma,\mu} z^{\mu}$$

are analytic in some subdomain V of U containing p_0 and

$$\Sigma'^{\infty}_{\sigma=-\infty} |h_{\sigma}(z)|^2$$

converges uniformly in V. The functions $x_{\pm\sigma} = h_{\pm\sigma}(z)$ map V continuously and isometrically into what might be called a generalized Minkowski space H of infinite complex dimension, with coordinates $x_{\pm\sigma}$, with the topology of Hilbert space for the purpose of defining continuity, and with the fundamental form $ds^2 = \Sigma'_{\sigma} (\operatorname{sgn}\sigma) |dx_{\sigma}|^2$ and corresponding diastasis

$$d(p,q) = \Sigma'_{\sigma}(\operatorname{sgn}\sigma) \left|x_{\sigma}(q) - x_{\sigma}(p)\right|^2$$

inducing metric and diastasis in V. In H we view $D(p,q)$ as an hermitian form on the vectors with components $x_{\sigma}(q) - x_{\sigma}(p)$ or as a real quadratic form on the real components of such vectors.

From the theory of real quadratic forms we use the following statements:

a) Given a real linear subspace H' of H spanned linearly by a set E, the form $D(p,q)$ restricted to H' is positive semi-definite of rank K if and only if for any finite set of points $p_0, p_1, \ldots, p_r \in E$ the symmetric matrix whose components are $\Delta_{ij} = D(p_0,p_i) + D(p_0,p_j) - D(p_i,p_j)$ is positive semi-definite and of rank $\leq \min(r,K)$, where equality can be achieved for every value of r for a suitable choice of points.

b) If H' is a complex linear subspace of H in which the diastasis is positive semi-definite and of rank K, then H' is uniquely a direct sum of two complex linear, mutually orthogonal subspaces L and H^0 (if we consider H as a Hilbert space) and such that $D(p,q) = 0$ for $p,q \in H^0$, and $D(p,q)$ defines a positive definite hermitian form on L; then L is a $K/2$-dimensional unitary space and the orthogonal projection of H' onto L along H^0 preserves the diastasis.

We now suppose V mapped into H such that p_0 is at the origin. Let W be a subdomain of V, at all of whose points the metric is resolvable of rank N. Let H' be the real linear subspace of H spanned by W, and hence by all of V. Then H' is complex linear and for all finite sets of $r+1$ points in W the matrix considered in statement a) is positive semi-definite of rank $\leq \min(r, 2N)$. Hence, by statement b) H' can be projected complex analytically onto a linear N-dimensional complex unitary subspace L preserving the diastasis. So all of V can be isometrically and complex analytically mapped onto L, whose completion is C^N. Hence the theorem is proved.

It is easy now to apply the last two theorems, and the monodromy principle to verify that if the metric on $\mathfrak{M}$ is analytic everywhere and resolvable of rank N at just one point, then the whole universal covering space $\widetilde{\mathfrak{M}}$ of $\mathfrak{M}$ can be globally immersed isometrically in C^N. The fundamental group of $\mathfrak{M}$ has thus a concrete representation into the group of motions of C^N. It is also to be noted that the diastasis $D(p,q)$ is then definable and real analytic in all pairs of points in the whole covering surface $\widetilde{\mathfrak{M}}$; it is everywhere ≥ 0 and becomes zero only for those pairs of points $\widetilde{\mathfrak{M}}$ that have a common image in C^N under the map. If $\tilde{p}$ is any point of $\widetilde{\mathfrak{M}}$, if σ is an element of the fundamental group $\pi_1(\mathfrak{M})$ of $\mathfrak{M}$, and $\tilde{p}^{\sigma}$ the translation of $\tilde{p}$ by σ, then the imbedding, considered as a multiple-valued map of $\mathfrak{M}$ into C^N, is single-valued, if, and only if $D(\tilde{p}, \tilde{p}^{\sigma}) = 0$ for all $\tilde{p}$ in a fundamental domain of $\mathfrak{M}$ in $\widetilde{\mathfrak{M}}$ and all generators σ of $\pi_1(\mathfrak{M})$. In this case $D(p,q)$ is single valued in $\mathfrak{M} \times \mathfrak{M}$. The map is a topological imbedding if, in addition, $D(p,q) = 0$ implies $p = q$. We thus arrive at our main theorem.

THEOREM 4. A metric Riemann surface can be isometrically and complex analytically imbedded in C^N if, and only if the metric is real analytic and resolvable of rank at least N at one point, and if the diastasis is everywhere single-valued. The mapping is a topological imbedding, provided that the diastasis between distinct points is positive.

4. REMARKS

There are some relations between the densities G_{μ} of Theorem 1 and the curvatures $\rho_1, \ldots, \rho_{N-1}$ at each point of $\mathfrak{M}$. In fact if $x_{\sigma} = f_{\sigma}(z)$ represent an isometric map of $\mathfrak{M}$ into C^N, we can verify that

$$G_{\mu}(z,\bar{z}) = \sum_{1 \leq \sigma_1 < \sigma_2 < \ldots < \sigma_{\mu} \leq N} \begin{vmatrix} f'_{\sigma_1}(z) & f'_{\sigma_2}(z) & \ldots & f'_{\sigma_{\mu}}(z) \\ f''_{\sigma_1}(z) & f''_{\sigma_2}(z) & \ldots & f''_{\sigma_{\mu}}(z) \\ \vdots & \vdots & & \\ f^{(\mu)}_{\sigma_1}(z) & f^{(\mu)}_{\sigma_2}(z) & \ldots & f^{(\mu)}_{\sigma_{\mu}}(z) \end{vmatrix}^2 = (g(z,\bar{z}))^{\frac{\mu(\mu+1)}{2}} \prod_{\nu=1}^{\mu-1} \rho_{\nu}^{2(\mu-\nu)}$$

From this we have

$$\frac{G_{\mu+1}(z,\overline{z})}{G_{\mu}(z,\overline{z})} = g^{\mu+1}\rho_1^2\rho_2^2\cdots\rho_\mu^2; \qquad \rho_\mu^{\,2} = \frac{G_{\mu+1}\,G_{\mu-1}}{g\,G_\mu^{\,2}}\ .$$

From (3) we obtain the recursive formula for the curvatures, using Δ to denote the operator

$$\frac{1}{g}\,\frac{\partial^2}{\partial_z\partial\,\overline{z}}\ ,$$

$$\rho_\mu^2 = \Delta\log\left[g^{\frac{\mu(\mu+1)}{2}}\rho_1^{\,2(\mu-1)}\rho_2^{\,2(\mu-2)}\cdots\rho_{\mu-1}^2\right].$$

Thus we have proved that all the curvatures of $\mathfrak{M}$ are computable from the metric. It is perhaps curious to note that $-\rho_1^{\,2}$ is The Gaussian curvature of $\mathfrak{M}$.

The methods outlined above can be directly applied to write criteria for any real analytic density of weight w to be developable into a series of squares of absolute values of holomorphic densities of weight w (including, of course, the case of scalars) or even to solve other functional equations of similar nature. The direct applications of these methods furnish new proofs of some specific statements, such as the follows. The hyperbolic (Poincaré) plane can be imbedded in Hilbert space and has complete mobility in itself there (originally proved by Bieberbach, [1]), but cannot be imbedded in a finite dimensional unitary space (cfr. Bochner, [2]). The generalized Minkowski space H is the "universal enveloping space" into which every metric Riemann surface can be locally imbedded. However, the map here is not uniquely determined (i.e. up to a diastasis-preserving transformation of H). In particular all algebraic curves with the metric induced from an elliptic Fubini (complex projective) space can be imbedded in H except for a hyperplane section of the curve.

---

1 This paper was prepared while the author held a fellowship from the Office of Naval Research, Project NR 045050.

2 The author has been shown an elegant counterexample by B. Lepson, in which the interior of the unit circle $|z| < 1$ is mapped complex analytically in Hilbert space, so that the map is discontinuous at every point where z is real.

REFERENCES

[1] BIEBERBACH, L., Eine singularitätenfreie Fläche konstanter negativer Krümmung im Hilbertschen Raum, Comm. Math. Helv. 4 (1932), pp. 248-255.

[2] BOCHNER, S., Curvature in Hermitian Metric, Bull. Amer. Math. Soc., 53, (1947), pp. 179-195.

[3] CALABI, E., Isometric, Complex Analytic Imbedding of Kähler Manifolds, Princeton Thesis, 1950. (Ann. of Math., 54 (1953), to appear)

[4] KLEIN, F., Über Riemanns Theorie der algebraischen Funktionen und ihrer Integrale, Leipzig, 1882.

SOME RESULTS RELATED TO EXTREMAL LENGTH

JAMES A. JENKINS

1. The method of extremal length and modifications of it serve to prove almost all geometric results in the theory of conformal mapping. Very frequently there is an analogous result for functions which need not be univalent but otherwise satisfy the conditions imposed on the function performing the conformal mapping. In certain simple cases such results can be obtained by using the principle of subordination. However in many other cases, usually those dealing with multiply-connected domains, this is no longer effective. The object of this report is to indicate circumstances under which it is possible to modify the method of extremal length to make it applicable to problems involving functions which are not univalent.

2. The general concept of extremal length is as follows. Let Ω be a domain of the z-plane and Γ a family of rectifiable curves contained in Ω. Rectifiability is to be understood in a local sense. Consider the class C of non-negative functions $\rho(z)$ defined on Ω such that $\int_\gamma \rho|dz|$ is defined (possibly $+\infty$) for all $\gamma \in \Gamma$ and such that

$$L_\rho(\Gamma) = \underset{\gamma \in \Gamma}{\text{g. l. b.}} \int_\gamma \rho|dx|$$

$$A_\rho(\Omega) = \iint_\Omega \rho^2 \, dxdy$$

are not simultaneously 0 or ∞. Provided the class C is non-void we define

$$\lambda(\Gamma) = \underset{\rho \in C}{\text{l. u. b}} \frac{L_\rho^2(\Gamma)}{A_\rho(\Omega)}$$

to be the extremal length of the family Γ.

It is readily seen that this definition is conformally invariant in the sense that any conformal mapping of Ω onto a domain Ω' will transform Γ into a family Γ' in Ω' with $\lambda(\Gamma') = \lambda(\Gamma)$. Actually the value of $\lambda(\Gamma)$ is independent of the domain Ω enclosing Γ but usually the family Γ is defined by some relationship to Ω. The definition can be at once extended to Riemann surfaces (since local rectifiability is conformally invariant) provided we consider instead of functions $\rho(z)$, conformal metrics $\rho(z)|dz|$ in the local uniformizing parameters z which transform in such a way that $\rho\,|dz|$ is invariant.

It is convenient to introduce two variants of the above definition. We will suppose that the class C is not void. First consider only those functions ρ for which

$$L_\rho(\Gamma) \geq 1.$$

We will call this the L-normalization of the problem Under this condition $\underset{\rho \in C}{\text{g. l. b.}}\, A_\rho(\Omega)$ is equal to $1/\lambda(\Gamma)$. This quantity will be called the module of the given configuration. Secondly consider only those functions ρ for which

$$A_\rho(\Omega) \leq 1.$$

We will call this the A-normalization of the problem. Then $\underset{\rho \in C}{\text{l. u. b.}}\, L_\rho(\Gamma)$ is equal to $(\lambda(\Gamma))^{1/2}$. Each of these two quantities varies monotonically with the extremal length.

The following are well known important examples of extremal lengths:

(i) Ω a rectangle with sides of lengths a and b, Γ the family of curves joining the sides of length a. Then $\lambda(\Gamma) = b/a$.

(ii) Ω the circular ring $r < |z| < R$, Γ the family of closed curves separating the two contours. Then

$$\lambda(\Gamma) = 2\pi/\log(R/r).$$

(iii) Ω the circular ring $r < |z| < R$, Γ the family of curves joining the two contours. Then

$$\lambda(\Gamma) = (1/2\pi)\log R/r$$

Evidently similar problems can be considered in conformally equivalent domains. Then curves running to the boundary must be allowed to tend to a prime end.

The examples given have the following properties in common:

(a) there is in each case an extremal metric i.e. a metric $\rho(z)|dz|$ for which $L_\rho(\Gamma)^2/A_\rho(\Omega)$ attains its least upper bound. Owing to the homogeneity of this expression, ρ is determined only up to a constant factor. If we eliminate this freedom by a normalization it is well known that whenever an extremal metric exists it is unique. In particular for the L-normalization the functions ρ are respectively (i) $1/b$, (ii) $(2\pi|z|)^{-1}$, (iii) $((\log R/r)|z|)^{-1}$.

(b) In each case there is a 1-parameter subfamily of Γ consisting (for the L-normalization) of curves whose exact length in the extremal metric is 1 and which sweep out the domain Ω i.e. through every point there is one and only one curve of the subfamily. We will call these the basic curves of the problem. They are respectively (i) parallels to the sides of length b, (ii) circles centre the origin, (iii) radial segments.

(c) the orthogonal trajectories of these curves also sweep out the domain Ω. We will call these the associated curves of the problem. They are respectively (i) parallels to the sides of length a, (ii) radial segments, (iii) circles centre the origin.

Length measured in the extremal metric along either one of these sets of curves induces in a natural manner a measure on the other set of curves.

3. As a first example of the method to be presented let us consider a problem treated by Ahlfors [1] and Polya [4]. Let R be the closed rectangle $-\chi \leq x \leq \chi$, $-m \leq y \leq m$, $z = x + iy$. The result to be proved is that there is no function $f(z)$ regular on R and satisfying the conditions:

(i) on the horizontal sides $|\mathcal{I} f(z)| \leq m$

(ii) on one vertical side $\mathcal{R} f(z) \leq -\chi$, on the other $\mathcal{R} f(z) \geq \chi$

other than the function $f(z) \equiv z$.

If the function $f(z)$ were required further to be univalent the result would follow immediately by the consideration of a suitable extremal length problem. On the one hand we can consider the problem of the extremal length of curves lying in R and joining the two vertical sides using the L-normalization. The extremal metric is given by the function $\rho = 1/2\chi$. We consider the corresponding problem in R^*, the image of R by the function $f(z)$. By conformal invariance the extremal metric is the one obtained from ρ by the mapping $f(z)$. We will assume that R^* lies in the same plane as R. Taking in R^* the metric corresponding to the function ρ' defined by

$$\rho' = 1/2\chi \quad \text{in } R^* \cap R$$
$$\rho' = 0 \quad \text{in } R^* - R,$$

we see that it is admissible in the problem for R^*. Indeed, as a consequence of conditions (i) and (ii) every curve of R^* joining the images of the vertical sides of R has, as a subcurve, a curve lying in R and joining the vertical sides of R, which thus has length at least 1 in the given metric. Now in this metric

$$A_{\rho'}(R^*) \leq A_{\rho}(R)$$

hence ρ' must give the extremal metric in R^*. Using the characterization of this extremal metric given above a familiar argument shows that we must have $f(z) \equiv z$.

Alternatively the result in this special case can be proved by considering the problem of the extremal length of curves lying in R and joining the two horizontal sides, using now the A-normalization. Here the extremal metric is given by the function $\rho = (4\chi m)^{-1/2}$. In it the above curves have length at least $(m/\chi)^{1/2}$. We consider the corresponding problem in R^*, the image of R by $f(z)$. The extremal metric is the one obtained from ρ by the mapping $f(z)$. We denote it by $\rho^*(z)$ (again assuming that R^* lies in the z-plane). Taking in R the metric corresponding to the function ρ' defined by

$$\rho' = \rho^* \quad \text{in } R^* \cap R$$
$$\rho' = 0 \quad \text{in } R - R^*$$

we see that it is admissible in the problem for R. Indeed $A_{\rho'}(R) \leq A_{\rho^*}(R^*) = 1$. In this metric, as a consequence of conditions (i) and (ii) every curve lying in R and joining the horizontal sides of R has, as a subcurve, a curve lying in R^* and joining the images of the horizontal sides of R. It thus has length at least $(m/\chi)^{1/2}$ in the given metric. The latter must therefore be the extremal metric for the problem in R and as before we deduce $f(z) \equiv z$.

Both of these procedures can be extended to the case where $f(z)$ need not be univalent. In this general situation the image of R will be a Riemann domain which we again denote by R^* and suppose to lie over the z-plane. At every point of R^* other than branch points we can use the local uniformizing parameter induced by the covering of the z-plane. Since there are here only a finite number of branch points we need not concern ourselves with them in defining a metric. Thus we will always understand that we are using these particular local uniformizing parameters.

In the first method our object is to obtain from the extremal metric on R for the problem considered an admissible metric for the corresponding problem on R^* in which the area of R^* does not exceed the area of R in its extremal metric. This is attained by constructing a simple covering in the following manner. Let us consider a basic curve of the problem for R and its image L^* on R^* under $f(z)$. Its projection L on the z-plane is an analytic arc (not necessarily simple) which meets at least once each of the associated curves of the problem for R. As we describe L we regard the points on R^* lying above the points where L meets a given associated curve A. Through each such point we draw a maximal segment on R^* lying over A. It is well known that at least one of these segments must join the images of the two horizontal sides of R and that this may be chosen to vary continuously with the point on L^* through which it passes apart from a finite number of exceptions. Subject to this condition we choose one such segment on R^* over each associated curve and define a metric $\rho'|dz|$ on R^* by setting $\rho' = 1/2\chi$ on these segments and $\rho' = 0$ elsewhere on R^*. Every curve on R^* joining the images of the vertical sides of R must cross each of these segments and thus have length at least 1 in this metric. The latter is therefore an admissible metric in the problem on R^*. Since the projection on the z-plane of the points of R^* where $\rho' \neq 0$ covers a subset of R simply, it is clear that $A_{\rho'}(R^*) \leq A_{\rho}(R)$. This means that $\rho'|dz|$

is the extremal metric on R^*. By conformal invariance it must also be the metric we obtain by conformal transfer of the extremal metric on R. In particular it can be nowhere zero. Thus R^* must lie simply over the z-plane and $f(z)$ must be univalent. The proof is now completed as before.

In the second method we begin with the extremal length problem for the family Γ^* of curves on R^* joining the images of the horizontal sides of R, using the A-normalization. Starting with the extremal metric for this problem, our object is to obtain an admissible metric for the corresponding problem on R in which each curve joining the horizontal sides of R has length at least equal to the minimal length occurring in the extremal metric. Let $\rho^*(z)|dz|$ denote the extremal metric on R^* where we use exclusively the local uniformizing parameters induced by covering the z-plane. Every point of the rectangle R is covered by a finite number (possibly zero) of points of R^*. Let $\rho'(z) = \max \rho^*(z)$ where $\rho^*(z)$ runs over the values at the points of R^* above z. In case there are no such points we define $\rho'(z) = 0$. The function $\rho'(z)$ gives an admissible metric in the problem considered in R. Indeed it is measurable on R and thus $A_{\rho'}(R) \leq A_{\rho^*}(R^*) = 1$. Now let γ be a curve in R joining the horizontal sides. Consider the set of curves lying on R^* and lying above γ. At least one such must join the images of the horizontal sides of R. It must then have length in the extremal metric at least $L_{\rho^*}(\Gamma^*)$ i.e., the minimal length of such curves in the extremal metric. However by the way ρ' was defined $\int_{\gamma_1} \rho'|dz|$ must also have at least this value. Hence the metric given by ρ' must be the extremal metric in R. Thus R^* must lie simply above R and $f(z)$ must be univalent. As before this gives $f(z) \equiv z$.

The assumption that $f(z)$ was analytic on the perimeter of the rectangle was made only for simplicity to make clear the essential features of the methods. Both methods apply equally well if it is dropped. We only have to observe that now there may be countably many branch points and that countably many points of the Riemann surface may lie above a given point of the domain. The auxiliary metric $\rho'|dz|$ can be defined in the same way as before in each case and in the same way we can verify that there are satisfied both the normalization inequality and the inequality which implies that the auxiliary metric must be extremal provided it is admissible. By this last qualification we mean that to use directly the definition of extremal length as given in §2 it is necessary to verify that $\int_{\gamma} \rho'|dz|$ is defined for all $\gamma \in \Gamma$. This presents no difficulty in the present simple situation if we assume the metrics defined only on the open rectangle and its image. In any case this question can always be avoided by the use of the following easily proven lemma:

LEMMA 1. Let an extremal length problem

satisfy conditions (a), (b), (c) of §2. If $\rho' \in L^2(\Omega)$ $\int_\gamma \rho' |dz|$ is defined for almost all basic curves γ. If for almost all such $\int_\gamma \rho' |dz| \geq 1$ then $A_{\rho'}(\Omega)$ is greater than or equal to the module of the problem and equal only when ρ' is almost everywhere equal to the extremal function.

The lemma is here stated to deal with the L-normalization. An analogous statement can be made for the A-normalization.

4. As a second example let us consider a problem first treated by M. Schiffer [5]. Let D be a doubly-connected domain. We will denote by $M(D)$ the module for the extremal length problem for the family of curves separating the two boundary continua, taken in the L-normalization. Every doubly-connected domain can be mapped on a circular ring and if the latter has radii r, R $(R > r)$ then $M(D) = (1/2\pi)\log R/r$ (see §2, example (ii)). Now suppose that a second domain D' can be mapped into D by a regular (not necessarily univalent) function $f(z)$ in a non-trivial manner (i.e. not homotopic to a constant mapping in D). Then Schiffer's result states that $M(D') \leq M(D)$. For convenience we may take D to be $r < |w| < R$, D' to be $r' < |z| < R'$ and the mapping to be given by $w = f(z)$, this function being regular for $r' \leq |z| \leq R'$. The topological condition above is equivalent to having the variation $\Delta \arg f(z)$ different from zero on one (and thus all) of the basic curves in the problem for D' corresponding to that stated above for D. The first procedure of §3 can now be applied to construct from the extremal metric on D and admissible metric on D' (satisfying the L-normalization condition) in which the area of D' is at most the minimal area of D. For this we use the projection of the image of a basic curve in the problem for D' in the same way as we used the curve L in §3. The associated curves are now radial segments. The result $M(D') \leq M(D)$ is an immediate consequence. If the degree of the image curve about $w = 0$ is precisely $\pm n$ $(n > 0)$ we can in an analogous manner obtain instead of a simple covering an n-fold covering. This will define on the Riemann surface which is the image of D' by $f(z)$ a metric $\rho'|dz|$ for which $L_{\rho'}(\Gamma') \geq n$, $A_{\rho'}(D') \leq nA_\rho(D)$ where ρ defines the extremal metric for the extremal length problem for D. This implies at once $nM(D') \leq M(D)$. It is easily established that equality is attained only in the case of a precise n-fold covering. This result can also be deduced from the first by the use of an auxiliary n^{th} root transformation. It can be used to simplify greatly some of the considerations in a recent paper by Huber [2].

The same results can be obtained by applying the second procedure of §3 to the problem of the extremal length of curves joining the two contours of each domain. As before the assumption that $f(z)$ is analytic on the boundary of D' is completely inessential.

As Schiffer has pointed out, this result can frequently be used in a manner analogous to the principle of subordination. In fact Schwarz's lemma can be proved by a slight modification of the above reasoning.

It should be pointed out that our proof actually gives appreciably more than the result as presented by Schiffer. Suppose in particular that the degree of the image curve above is positive. Then we need not assume that the image of D' lies in D but only that it lies in $|z| < R$ and that the image of its inner contour lies in D. This is easily verified by checking through the details of the proof. This extension is sometimes quite useful.

5. The result of the preceding section can be extended also to domains of higher connectivity. One such extension is as follows. Let D be a domain, n-fold-connected, lying in the z-plane $(z = x + iy)$ and for convenience assumed to be bounded by analytic curves. We distinguish one of its boundary curves which we denote by C. Suppose that there is a function $f(z)$ regular on $\overline{D}$ the closure of D, taking values in $\overline{D}$ and taking C into a curve homotopic on $\overline{D}$ to C or to C traced a finite number of times. The statement is that this is possible only if $f(z)$ maps D conformally (i.e. (1,1)) onto itself. This result was indicated to me by Schiffer who gave quite a different proof.

In order to apply the first procedure of §3 it is necessary to use a generalization of extremal length. Instead of a family of individual curves we use a family Γ each element of which is a set of a finite number of curves separating C from the remaining boundaries of D. With these we proceed as before, using the L-normalization. We consider metrics $\rho(z)|dz|$ in which the total length of such a set of curves is always at least 1. Then we ask for the least possible area of D in such metrics. It is now to be shown that this problem actually has an extremal metric.

This metric is obtained by considering the harmonic measure $\omega(z)$ of C with respect to D i.e. that harmonic function defined on D which has the value 1 on C and the value 0 on the other boundaries of D. In the metric $\rho^*(z)|dz|$ where $\rho^* = ((\frac{\partial\omega}{\partial x})^2 + (\frac{\partial\omega}{\partial y})^2)^{1/2}$ each set of curves belonging to Γ has length at least $|\int_C \frac{\partial\omega}{\partial n} ds|$. The sets of level curves of ω at a fixed level play the role of basic curves. Each set has precisely the given total length and they sweep out the domain D except for the presence of a finite number of exceptional points, namely the critical points of ω. These cause no difficulty and thus it is easily seen by standard methods that the metric $\rho(z)|dz|$, where $\rho = |\int_C \frac{\partial\omega}{\partial n} ds|^{-1}((\frac{\partial\omega}{\partial x})^2 + (\frac{\partial\omega}{\partial y})^2)^{1/2}$, is the desired extremal metric. We observe the further important fact that the orthogonal trajectories of these curves run from C to one of the other boundaries of D with the exception of a finite number of curves which end at the critical points of ω. Clearly the presence of a finite number of such exceptional curves offers no obstacle to our method. Thus we can apply our first procedure of §3 to D^*, the image of D by $f(z)$, using the image C^* of C in the role of the curve L^*. This shows that D^* must coincide with D so that $f(z)$ must be a conformal mapping of D onto itself as stated.

The assumption concerning analyticity of $f(z)$ on the boundary of D is again unessential. In place of C^* one uses the image of a basic curve of the problem isotopic to C. Naturally the method can also be applied to two distinct domains to obtain inequalities for the modules in question.

The result has a generalization analogous to that given in the last paragraph of §4. Using a slight variant of this we can obtain the Aumann-Carathéodory Starrheitssatz for multiply-connected domains.

The result of this section can likewise be obtained using the second procedure of §3. Here one takes the extremal length of the family of curves joining C to any of the other boundaries of D, using the A-normalization. The extremal metric is again derived from the harmonic measure ω of C. The basic curves now run from C to the other boundaries of D (with the exception of a finite number which meet the critical points of ω.)

6. The methods presented in this report can be used to treat many problems besides those of the preceding sections. They can be used to prove Löwner's Lemma and certain generalizations thereof. They can be used to give a rather simple proof of Picard's Second Theorem. Let us also remark that with appropriate changes all results obtained by this method apply also to functions related to regular functions in the same way that quasiconformal mappings are related to conformal mappings.

Finally these methods can be used to simplify the proofs of results occurring throughout the literature. In this connection let us mention only a paper of several years ago by Nehari [3].

BIBLIOGRAPHY

[1] AHLFORS, L., Sur une généralisation du théorème de Picard, C. R. Acad Paris, 194 (1932), pp. 245-247.

[2] HUBER, H., Über analytische Abbildungen von Ringgebieten in Ringgebieten: Comp. Math., 9 (1951), pp. 161-168.

[3] NEHARI, Z., On analytic functions possessing certain properties of univalency. Proc. London Math. Soc. 50 (1949), pp. 120-136.

[4] POLYA, G., Über analytische Deformationen eines Rechtecks, Ann. of Math. 34 (1933), pp. 617-620.

[5] SCHIFFER, M., On the modulus of doubly-connected domains, Quart. Jour. Math. 17 (1946), pp. 197-213.

RANDOM WALK AND THE TYPE PROBLEM OF RIEMANN SURFACES

SHIZUO KAKUTANI

§1. INTRODUCTION

Let Σ^2 be a simply connected non-compact two-dimensional Riemann surface. Then, by a fundamental theorem in the theory of conformal mapping, there exists an analytic function defined on Σ^2 which maps Σ^2 conformally onto the entire plane or onto the interior of the unit circle. We say that the type of Σ^2 is <u>parabolic</u> in the first case, and <u>hyperbolic</u> in the second case. One of the most interesting problems in the theory of Riemann surfaces is to determine the type of a given Riemann surface. Our main purpose is to give a criterion for this type problem in terms of random walks on a Riemann surface.

It is well known that random walks in an n-dimensional real Euclidean space R^n have entirely different properties in the cases $n = 2$ and $n \geqq 3$. (Cf. G. Pólya [9], A. Dvoretzky and P. Erdös [2], K. L. Chung and W. H. J. Fuchs [1]). Let us consider, for example, a discrete random walk in R^n starting from the origin of R^n in which each step consists in walking a unit length in one of $2n$ directions parallel to the positive and negative directions of n coordinate axes. Then, almost all orbits in R^2 are recurrent and travel through all integer points of R^2 (integer points are those points whose coordinates are all integers), while in the case of R^n with $n \geqq 3$, almost all orbits diverge to infinity (without travelling through all integer points of R^n) as the number of steps increases to infinity.

Similar phenomena can be observed in the case of a continuous random walk in R^n, for example, a Brownian motion in R^n. Almost all paths of a Brownian motion in R^2 are everywhere dense in R^2 and come back to any neighborhood of any point of R^2 infinitely many times for arbitrary large values of time, while almost all paths of a Brownian motion in R^n with $n \geqq 3$ are nowhere dense in R^n and diverge to infinity (i.e. the distance from the origin of R^n tends to infinity) as time increases to infinity. (Cf. S. Kakutani [3]).

If we, however, consider random walks on a Riemann surface, then the situation is entirely different. To make the arguments simpler, we discuss in this paper only the cases when an n-dimensional Riemann surface Σ^n is given as a covering surfave over R^n. This means that there exists a continuous mapping $z = \varphi(\zeta)$ of Σ^n into R^n which is a local homeomorphism. Then, since all paths of a Brownian motion in R^n are continuous curves, it is possible to define a Brownian motion on Σ^n as the continuous inverse image of a Brownian motion in R^n by the inverse mapping $\zeta = \varphi^{-1}(z)$. It is possible to show that one of the following mutually exclusive cases happens: (I) almost all paths of a Brownian motion on Σ^n are everywhere dense in Σ^n and come back to any neighborhood of any point of Σ^n infinitely many times for arbitrary large values of time; (II) almost all paths of a Brownian motion on Σ^n are nowhere dense in Σ^n and diverge to the "boundary" of Σ^n in the sense that, for any compact subset F of Σ^n, each of these paths stays outside of F after a certain value of time. We say that Σ^n is of type (I) in the first case, and of type (II) in the second case.

It is easy to see that all Riemann surfaces Σ^n over R^n with $n \geqq 3$ are of type (II) since the underlying space R^n is of type (II) for $n \geqq 3$. But there are Riemann surfaces Σ^2 over R^2 of each type. It turns out that, in case $n = 2$ and Σ^2 is simply connected, (I) happens if and only if Σ^2 is parabolic, and (II) happens if and only if Σ^2 is hyperbolic. This fact makes it possible to find a criterion for the type problem of Riemann surfaces in terms of Brownian motion on such Riemann surfaces. (Cf. S. Kakutani [5]).

In case Σ^n is given as an abstract Riemann surface and not as a covering surface over R^n, the situation is more complicated. We can still define Brownian motions on such a Riemann surface Σ^n in a similar way, and can show that one of the cases (I) or (II) will happen. But the type of Σ^n does not depend on the dimension of Σ^n (in fact, for each dimension $n \geqq 2$, there exist Riemann surfaces Σ^n of each type), and is determined by the "metric structure" of Σ^n near its "boundary".

In §2 we state some results on Brownian motions in R^n which are needed in the discussions in §3. In §3 we give the formulation of our main results. These results are stated only for Riemann surfaces which are given as covering surfaces over R^n, but our main theorem holds also for general Riemann surfaces. The details of proofs will be given in a subsequent paper.

It is possible to discuss discrete random walks on a Riemann surface which are very similar to the Markoff process in R^2 previously discussed by the author [6] in connection with the Dirichlet problem. We can obtain the same classification of Riemann surfaces by means of the asymptotic behavior of such discrete random walks on these Riemann surfaces. The details will be discussed on another occasion.

§2. BROWNIAN MOTION IN n-SPACE R^n

Let $(\Omega, \mathcal{E}, Pr)$ be a probability space, i.e. $\Omega = \{\omega\}$ is a set of elements ω, $\mathcal{E} = \{E\}$ is a σ-field of subsets E of Ω, and $Pr(E)$

is a countably additive set function defined on $\mathcal{E}$ with $\Pr(\Omega) = 1$.

Let $R^n = \{z\}$ be an n-dimensional real Euclidean space (or simply, an n-space), and let $z(t,\omega)$ be an R^n-valued function of two variables t and ω defined for $0 \leqq t < \infty$ and $\omega \in \Omega$. $z(t,\omega)$ is called a Brownian motion in R^n if the following conditions are satisfied: (i) $z(0,\omega) = 0$ for all $\omega \in \Omega$; (ii) for any $t > 0$, $z(t,\omega)$ is a measurable function of ω and has an n-dimensional Gaussian distribution which is given by the formula:

$$\Pr\left\{\omega \mid z(t,\omega) \in B\right\} = \left(\frac{1}{2\pi t}\right)^{\frac{n}{2}} \int \cdots \int_B \exp\left(-\frac{u_1^2 + \cdots + u_n^2}{2t}\right) du_1 \cdots du_n \tag{2.1}$$

for any Borel subset B of R^n; (iii) for any s_k, t_k, $k = 1,\ldots,p$ with $0 \leqq s_1 < t_1 \leqq s_2 < t_2 \leqq \cdots \leqq s_p < t_p < \infty$, $\left\{z(t_k,\omega) - z(s_k,\omega) \mid k = 1,\ldots,p\right\}$ is an independent system in the sense of probability, i.e.

$$\Pr\left\{\omega \mid z(t_k,\omega) - z(s_k,\omega) \in B_k, \quad k = 1,\ldots,p\right\} = \prod_{k=1}^{p} \Pr\left\{\omega \mid z(t_k,\omega) - z(s_k,\omega) \in B_k\right\} \tag{2.2}$$

for any Borel subsets $B_1,\ldots,B_p$ of R^n.

It was proved by N. Wiener [10] and P. Lévy [7,8] that, for any Brownian motion $z(t,\omega)$ in R^n, there exists a Brownian motion $z'(t,\omega)$ in R^n with the following properties: (i) for any $t \geqq 0$, $z'(t,\omega) = z(t,\omega)$ for almost all $\omega \in \Omega$; (ii) for any $\omega \in \Omega$, $z'(t,\omega)$ is a continuous function of t for $0 \leqq t < \infty$. Hence, in the following pages we shall assume that, for any $\omega \in \Omega$, $z(t,\omega)$ is a continuous function of t for $0 \leqq t < \infty$.

For any $z_0 \in R^n$, an R^n-valued function $z_0 + z(t,\omega)$ of two variables t and ω defined for $0 \leqq t < \infty$ and $\omega \in \Omega$ is called a Brownian motion in R^n starting from z_0. For any $\omega \in \Omega$, $C(z_0,\omega) = \left\{z_0 + z(t,\omega) \mid 0 \leqq t < \infty\right\}$ is called the ω-path of the Brownian motion $z_0 + z(t,\omega)$ in R^n starting from z_0.

Let F be a compact subset of R^n. Then, for any $z_0 \in R^n - F$, there exists a generalized real number $\tau = \tau(z_0,\omega, F, R^n)$ $(0 < \tau \leqq \infty)$ uniquely determined by the following properties: (i) $z_0 + z(t,\omega) \in R^n - F$ for $0 \leqq t < \tau$; (ii) τ is the largest generalized real number with the property (i). It is clear that $\tau = \infty$ means that the ω-path of the Brownian motion $z_0 + z(t,\omega)$ in R^n starting from z_0 never reaches F, and $\tau < \infty$ means that $\alpha(z_0,\omega, F, R^n) = z_0 + z(\tau,\omega)$ is the point in F where the ω-path of the Brownian motion $z_0 + z(t,\omega)$ in R^n starting from z_0 reaches F for the first time. It is easy to see that $\tau(z_0,\omega, F, R^n)$ is a generalized real-valued measurable function of ω

defined on Ω, and that $\alpha(z_o, \omega, F, R^n)$ is an R^n-valued measurable function of ω defined on the measurable set $\Omega(z_o, F, R^n) = \{ \omega \mid \tau(z_o, \omega, F, R^n) < \infty\}$.

Let us put

$$u(z_o, F, R^n) = Pr\left\{ \Omega(z_o, F, R^n) \right\}. \tag{2.3}$$

Then $u(z_o, F, R^n)$ is the probability that the Brownian motion $z_o+z(t,\omega)$ in R^n starting from z_o ever reaches F. The following result is known. (Cf. S. Kakutani [4]):

LEMMA 1. $u(z_o, F, R^n)$ is a harmonic function of z_o defined on $R^n - F$ with the following properties: (I) Case $n = 2$. $u(z_o, F, R^2) = 1$ for all $z_o \in R^2 - F$ if the logarithmic capacity of F is positive, and in particular, if F has an interior point; $u(z_o, F, R^n) \equiv 0$ for all $z_o \in R^2 - F$ if the logarithmic capacity of F is zero. (II) Case $n \geqq 3$. $u(z_o, F, R^n) = 1$ if and only if z_o belongs to a component of $R^n - F$ which is bounded. $u(z_o, F, R^n) < 1$ if and only if z_o belongs to the component of $R^n - F$ which extends to infinity and $u(z_o, F, R^n) \to 0$ as z_o tends to infinity. $u(z_o, F, R^n) = 0$ for some $z_o \in R^n - F$ if and only if the $(n-2)$-dimensional capacity of F is zero, in which case $u(z_o, F, R^n) = 0$ for all $z_o \in R^n - F$.

From this follows the following result (Cf. S. Kakutani [3]):

LEMMA 2. (I) Case $n = 2$. For any $z_o \in R^2$ and for almost all $\omega \in \Omega$, the ω-path of a Brownian motion in R^2 starting from z_o is everywhere dense in R^2 and comes back to any neighborhood of any point of R^2 infinitely many times for arbitrary large values of t (i.e. for any open subset O of R and for any t_o, there exists a $t > t_o$ such that $z_o + z(t,\omega) \in O$.) (II) Case $n \geqq 3$. For any $z_o \in R^n$ and for almost all $\omega \in \Omega$, the ω-path of a Brownian motion in R^n starting from z_o is nowhere dense in R^n and diverges to infinity as time tends to infinity (i.e. for any compact subset F of R^n, there exists a $t_o = t_o(z_o, \omega, F)$ such that $z_o + z(t,\omega) \in R^n - F$ for all $t > t_o$).

§3. BROWNIAN MOTION ON A RIEMANN SURFACE

Let Σ^n be a Riemann surface which is given as a <u>covering surface</u> over R^n, $n \geqq 2$. $\Sigma^n = \{\zeta\}$ is a connected topological space and there is an R^n-valued continuous function $z = \varphi(\zeta)$ defined on Σ^n with the following property: for any $\zeta_0 \in \Sigma^n$ there exists a neighborhood $V(\zeta_0)$ of ζ_0 in Σ^n which is mapped by $z = \varphi(\zeta)$ homeomorphically onto an n-dimensional open sphere in R^n with center at $\varphi(\zeta_0)$. $\varphi(\zeta)$ is called the <u>local coordinate function</u> on Σ^n. Let $C = \{z(t) \mid 0 \leqq t < \infty\}$ be a continuous curve in R^n, and let ζ_0 be a point of Σ^n such that $\varphi(\zeta_0) = z(0)$. Then there exists a generalized real number $\sigma = \sigma(\zeta_0, C, \Sigma^n)$ uniquely determined by the following properties: (i) there exists a continuous curve $\Gamma = \{\zeta(t) \mid 0 \leqq t < \sigma\}$ on Σ^n with the initial condition $\zeta(0) = \zeta_0$ such that $\varphi(\zeta(t)) = z(t)$ for $0 \leqq t < \sigma$; (ii) σ is the largest generalized real number with the property (i). $\zeta(t)$ may be considered as the branch of a many-valued function $\varphi^{-1}(z(t))$ with the initial condition $\zeta(0) = \zeta_0$, and $\sigma(\zeta_0, C, \Sigma^n)$ is the upper limit to which this branch can be continued as a continuous function of t. Γ is called the <u>maximal inverse image</u> of C with the initial point ζ_0.

Let Σ^n be a Riemann surface over R^n with the local coordinate function $\varphi(\zeta)$, and let $z(t,\omega)$ be a Brownian motion in R^n. For any $\zeta_0 \in \Sigma^n$ and for any $\omega \in \Omega$, consider the ω-path $C(\varphi(\zeta_0),\omega) = \{\varphi(\zeta_0) + z(t,\omega) \mid 0 \leqq t < \infty\}$ of the Brownian motion $\varphi(\zeta_0) + z(t,\omega)$ in R^n starting from $\varphi(\zeta_0)$. Let $\Gamma(\zeta_0,\omega) = \{\zeta(\zeta_0, t, \omega) \mid 0 \leqq t < \sigma(\zeta_0, \omega, \Sigma^n)\}$ be the maximal inverse image of $C(\varphi(\zeta_0), \omega)$ on Σ^n with the initial point ζ_0, where $\zeta(\zeta_0, t, \omega)$ is the branch of $\varphi^{-1}(\varphi(\zeta_0) + z(t,\omega))$ with the initial condition $\zeta(\zeta_0, 0, \omega) = \zeta_0$ defined for $0 \leqq t < \sigma(\zeta_0, \omega, \Sigma^n) = \sigma(\zeta_0, C(\varphi(\zeta_0),\omega), \Sigma^n)$. It is easy to see that $\sigma(\zeta_0, \omega, \Sigma^n)$ is a generalized real-valued measurable function of ω defined on Ω. For any $\zeta_0 \in \Sigma^n$, an Σ^n-valued function $\zeta(\zeta_0, t, \omega)$ of two variables t and ω defined for $0 \leqq t < \sigma(\zeta_0, \omega, \Sigma^n)$ and for all $\omega \in \Omega$, is called the <u>Brownian motion on Σ^n starting from</u> ζ_0 which corresponds to the Brownian motion $\varphi(\zeta_0) + z(t, \omega)$ in R^n starting from $\varphi(\zeta_0)$. Further, for any $\omega \in \Omega$, the continuous curve $\Gamma(\zeta_0, \omega) = \{\zeta(\zeta_0, t, \omega) \mid 0 \leqq t < \sigma(\zeta_0, \omega, \Sigma^n)\}$ on Σ^n is called the <u>ω-path</u> of the Brownian motion $\zeta(\zeta_0, t, \omega)$ on Σ^n starting from ζ_0.

Let F be a compact subset of Σ^n. For any $\zeta_0 \in \Sigma^n - F$, there exists a generalized real number $\tau = \tau(\zeta_0, \omega, F, \Sigma^n)$ $(0 < \tau \leqq \sigma(\zeta_0, \omega, \Sigma^n))$ uniquely determined by the following properties: (i) $\zeta(\zeta_0, t, \omega) \in \Sigma^n - F$ for $0 \leqq t < \tau$; (ii) τ is the largest generalized real number with the property (i). It is clear that $\tau = \sigma$ means that the ω-path of the Brownian motion $\zeta(\zeta_0, t, \omega)$ on Σ^n starting from ζ_0 never reaches F, and $\tau < \sigma$ means that $\alpha(\zeta_0, \omega, F, \Sigma^n) \equiv \zeta(\zeta_0, \tau, \omega)$ is the point in F

where the ω-path of the Brownian motion $\zeta(\zeta_0, t, \omega)$ starting from ζ_0 reaches F for the first time. It is easy to see that $\tau(\zeta_0, \omega, F, \Sigma^n)$ is a generalized real-valued measurable function of ω defined on Ω, and that $\alpha(\zeta_0, \omega, F, \Sigma^n)$ is an Σ^n-valued measurable function of ω defined on the measurable set $\Omega(\zeta_0, F, \Sigma^n) = \{\omega \mid \tau(\zeta_0, \omega, F, \Sigma^n) < \sigma(\zeta_0, \omega, \Sigma^n)\}$.

Let us put

$$(3.1) \qquad u(\zeta_0, F, \Sigma^n) = \Pr\left\{\Omega(\zeta_0, F, \Sigma^n)\right\}.$$

Then $u(\zeta_0, F, \Sigma^n)$ is the probability that the Brownian motion $\zeta(\zeta_0, t, \omega)$ on Σ^n starting from ζ_0 ever reaches F. $u(\zeta_0, F, \Sigma^n)$ is a harmonic function of ζ_0 defined on $\Sigma^n - F$, but it is not necessarily true that $u(\zeta_0, F, \Sigma^n) = 1$ for all $\zeta_0 \in \Sigma^n - F$ even when $n = 2$ and when F contains an interior point. We may summarize our main results as follows:

THEOREM 1. Let Σ^n be a Riemannian suface over R^n $(n \geq 2)$. Then one of the following two exclusive cases happens: (if $n = 2$, the (n-2)-dimensional capacity in the following statements should be replaced by the logarithmic capacity)

(I) $u(\zeta_0, F, \Sigma^n) = 1$ for all $\zeta_0 \in \Sigma^n - F$ if the (n-2)-dimensional capacity of F is positive, and in particular, if F has an interior point; $u(\zeta_0, F, \Sigma^n) = 0$ for all $\zeta_0 \in \Sigma^n - F$ if the (n-2)-dimensional capacity of F is zero. In this case, $\sigma(\zeta_0, \omega, \Sigma^n) = \infty$ for all $\zeta_0 \in \Sigma^n$ and for almost all $\omega \in \Omega$. Further, for any $\zeta_0 \in \Sigma^n$ and for almost all $\omega \in \Omega$, the ω-path of a Brownian motion $\zeta(\zeta_0, t, \omega)$ on Σ^n starting from ζ_0 is everywhere dense in Σ^n and comes back to any neighborhood of any point of Σ^n infinitely many times for arbitrary large values of t (i.e. for any open subset O of Σ^n and for any t_0 there exists a $t > t_0$ such that $\zeta(\zeta_0, t, \omega) \in O$).

(II) $u(\zeta_0, F, \Sigma^n) < 1$ if $\Sigma^n - F$ has only one unbounded component and if ζ_0 belongs to this component. (A subset of Σ^n is <u>bounded</u> if it is contained in a compact subset of Σ^n). If $\Sigma^n - F$ has only one unbounded component, then $u(\zeta_0, F, \Sigma^n) \to 0$ as ζ_0 tends to the "boundary" of Σ^n (i.e. for any $\varepsilon > 0$ there exists a compact subset F' of Σ^n such that $\zeta_0 \in \Sigma^n - F'$ implies $u(\zeta_0, F, \Sigma^n) < \varepsilon$). (In case $\Sigma^n - F$ has more than one unbounded components, the situation is more complicated). Further $u(\zeta_0, F, \Sigma^n) = 0$ for some $\zeta_0 \in \Sigma^n - F$ if and only if the (n-2)-dimensional capacity of F is zero, in which case $u(\zeta_0, F, \Sigma^n) \equiv 0$ for all $\zeta_0 \in \Sigma^n - F$. For any $\zeta_0 \in \Sigma^n$ and for almost all $\omega \in \Omega$, the ω-path of the Brownian motion $\zeta(\zeta_0, t, \omega)$ on Σ^n starting from ζ_0 is nowhere dense in Σ^n and tends to the "boundary" of Σ^n as t increases to $\sigma(\zeta_0, \omega, \Sigma^n)$. (The last statement means that for any compact subset F of Σ^n there exists a $t_0 = t_0(\zeta_0, \omega, F, \Sigma^n) < \sigma(\zeta_0, \omega, \Sigma^n)$ such that $t_0 < t < \sigma$ implies $\zeta(\zeta_0, t, \omega) \in \Sigma^n - F$).

BIBLIOGRAPHY

[1] CHUNG, K. L., FUCHS, W. H. J., On the distribution of values of sums of random variables, Memoires of Amer. Math. Soc. No. 4, Part I (1951).

[2] DVORETZKY, A., ERDÖS, P., Some problems on random walk in space, Proc. Second Berkeley Symposium of Math. Statistics and Probability, August 1950 (1951), 353-367.

[3] KAKUTANI, S., On Brownian motion in n-space, Proc. Acad. Japan, 20 (1944), 648-652.

[4] KAKUTANI, S., Two-dimensional Brownian motion and harmonic functions, Proc. Acad. Japan, 20 (1944) 708-714.

[5] KAKUTANI, S., Two-dimensional Brownian motion and the type problem of Riemann Surfaces, Proc. Acad. Japan, 21 (1945), 138-140.

[6] KAKUTANI, S., Markoff process and the Dirichlet problem, Proc. Acad. Japan, 21 (1945), 227-233.

[7] LÉVY, P., Théorie de l'addition des variables aléatoires, Paris, 1937.

[8] LÉVY, P., Processus stochastiques et mouvement brownien, Paris, 1948.

[9] PÓLYA, G., Über eine Aufgabe der Wahrscheinlichkeitsrechnung betreffend die Irrfahrt im Strassennetz, Math. Ann. 84 (1921), 149-160.

[10] WIENER, N., Generalized harmonic analysis, Acta Math. 55 (1930), 117-258.

CONSTRUCTION OF PARABOLIC RIEMANN SURFACES BY THE GENERAL REFLECTION PRINCIPLE

WILFRED KAPLAN

§1. INTRODUCTION

Let $w = f(z)$ be analytic in a domain D bounded by a single open curve C extending from infinity to infinity and let $v = \text{Im}[f(z)]$ have constant boundary values k everywhere on C. If C is a straight line, then the Schwarz reflection principle shows that $f(z)$ is an entire function. If C is a general curve, even analytic, no such conclusion follows. However, if D is mapped in a one-to-one conformal fashion on the upper half T of a complex t-plane, then the reflection principle shows that $w = f(z)$ becomes an entire function $h(t) = f[z(t)]$. Thus the original function in D can be regarded as a sort of "functional element" from which an entire function can be formed by symmetrization. It is easily seen that the symmetrization is equivalent to reflection in the line $v = k$ of the partial Riemann surface onto which $w = f(z)$ maps D. If the Riemann surface is represented by a planar graph G, the procedure yields a surface whose graph is obtained from G by an analogous symmetrization.

Particular functional elements $f(z)$ are obtainable, for example, from known entire functions $g(z)$; one need only choose D to be a domain bounded by a level curve of $v = \text{Im}[g(z)]$. In this way and by repetition of the process, a large class of entire functions can be generated; the functions obtained are in general unfamiliar, since the mapping of D on T is in general unfamiliar. However, the Riemann surfaces are obtained simply by reflection, as remarked above, and are hence explicitly known. Therefore the procedure should more properly be regarded as one for generating parabolic Riemann surfaces from pieces of known parabolic surfaces.

The method just described is actually only a special case of a much more general one for construction of parabolic (or, if desired, hyperbolic) Riemann surfaces. This will be made clear in the following sections, in which a number of examples are offered.

The results given here represent the initial stage of a program whose goal is the construction and classification of a large family of Riemann surfaces. Since the <u>type</u> of the surface will in general be known, the program can be regarded as a constructive approach to the type problem.

§2. HALF-PLANAR ELEMENTS

From the function $w = g(z) = z^2$ we can obtain four different kinds of elements, depending on the choice of D: D_1: the domain $y > x^{-1} > 0$; D_2: the interior of the complement of D_1; D_3: $0 < \arg z < 3\pi/2$; D_4: $y > 0$. The symmetrized functions $h(t)$ are respectively t, a third degree polynomial, t^3, t^2.

In general, if $g(z)$ is a polynomial, $h(t)$ will be a polynomial. It is of interest to ask what polynomials can be generated from $w = z^2$ by repeatedly applying the symmetrization process; this is equivalent to asking for the class of Riemann surfaces obtainable from that of $w^{1/2}$ by repeated reflection of "half-surfaces" in straight lines. This is being investigated by D. Wend.

From the function $w = g(z) = e^z$ three types of elements are obtained: D_1: the domain $e^x \sin y > 1$, $0 < y < \pi$; D_2: the interior of the complement of D_1; D_3: the half-plane $y > 0$. D_1 and D_3 give $h(t) = t$ and $h(t) = e^t$. D_2 gives a function $h(t)$ whose inverse has two logarithmic branch points; this can be shown to be (apart from linear transformations) the error function. By repeating the process, we obtain functions whose inverses have 3, 4, 5,... logarithmic branch points.

§3. STRIP ELEMENTS

We now choose D to be a domain bounded by two open curves C_1, C_2, both going from infinity to infinity, and $f(z)$ to be a function analytic in D whose imaginary part $v(z)$ has constant boundary values k_1, k_2 on C_1, C_2 respectively. The domain D can be mapped in one-to-one conformal fashion on an infinite strip in the t-plane. By repeated reflection in parallels to the edges of the strip, we obtain an entire function $h(t)$. The reflection can again be interpreted on the Riemann surface of the inverse function. If $k_1 = k_2$, $h(t)$ will be periodic; otherwise, only $h'(t)$ will be periodic.

Elements $f(z)$ can be obtained in domains bounded by two level curves of the imaginary part of an entire function $g(z)$. For example, from $g(z) = z^2$ we obtain an element in the domain $-1 < xy < 1$. The symmetrical function is found (apart from a linear transformation) to be $\sin t$.

From $w = g(z) = e^z$ we obtain a function $h(t)$ whose inverse has infinitely many logarithmic branch points by using as D the domain bounded by the curves $e^x \sin y = 1$, $0 < y < \pi$ and $e^x \sin y = -1$, $\pi < y < 2\pi$. By a similar operation on $h(t)$ we obtain a function whose inverse has logarithmic branch points which form a rectangular net over the w-plane.

A third application of the process in a well chosen manner yields a function whose inverse has logarithmic branch points which are everywhere dense over the w-plane. This illustrates the ease with which extremely complicated parabolic Riemann surfaces can be generated by reflections. It would be of interest to know whether the final function $h(t)$ obtained in this example is a "Gross function"; that is, a function whose inverse has every point of the w-plane as a singularity on some path.

§4. USE OF GENERAL REFLECTION PRINCIPLE

Let $w = f(z)$ be given as in the introduction and let the boundary curve C be analytic. The general reflection principle then implies that $f(z)$ can be continued across C. Under further assumptions, we can say much more about how far the continuation can be carried out:

> THEOREM 1. Let $w = f(z)$ be analytic in a simply-connected domain D and let $v = \mathrm{Im}[f(z)]$ have constant boundary values on an open analytic curve C which forms part of the boundary of D. Let there exist a one-to-one conformal mapping $w = \phi(z)$ of D on the upper half-plane which (on extension to the boundary) transforms C onto the entire real axis. Then $f(z)$ can be continued analytically wherever $\phi(z)$ can be continued. In particular, if $\phi(z)$ is an entire function, then $f(z)$ is entire.

This theorem can be proved in quite elementary fashion. The nature of the continuation can be shown by the following example. Let D be the domain $e^x \sin y > 1$, $0 < y < \pi$. Then $w = e^z = \phi(z)$ satisfies the conditions of the theorem. Accordingly, every function $f(z)$ which is analytic in D and whose imaginary part has constant boundary values k on the boundary of D is an entire function. The continuation of $f(z)$ to the finite z-plane can be accomplished as follows: for each z outside D and such that $z = z_0 + 2n\pi i$ for some z_0 in D plus boundary we set $f(z) = f(z_0)$; for every other z we find a unique z_1 in D such that e^z is the reflection of e^{z_1} in the line $v = 1$, and we let $f(z)$ be the reflection of $f(z_1)$ in the line $v = k$. Thus in a sense the entire function e^z has a "group" of automorphisms which we use to extend each function defined in the half-fundamental-domain D; these automorphisms have been studied recently by af Hällström [1] and others.

To obtain elements $f(z)$ we can choose a half-planar element $f_1(z)$ in D_1, as in Section 2, and map D_1 onto D.

The procedure can be described in terms of reflections of pieces of Riemann surfaces, but is in general much more complicated than those previously described.

If D is a subdomain of the unit circle, while $\phi(z)$ has $|z| = 1$ as natural boundary then "in general" $f(z)$ will also have $|z| = 1$ as natural boundary. This remark can be used as the basis for construction of hyperbolic Riemann surfaces. If, in particular, ϕ is automorphic (in the precise sense), then $f(z)$ is also automorphic.

BIBLIOGRAPHY

[1] HÄLLSTRÖM, af, G. Über die Automorphiefunktionen meromorpher Funktionen. Acta Acad. Aboensis XV No. 4 (1949) pp. 1-28.

ON THE IDEAL BOUNDARY OF A RIEMANN SURFACE

H. L. ROYDEN

In the study of a planar Riemann surface it is often convenient to consider it as a domain in the complex plane and draw conclusions about it from such properties of its complement as having zero capacity or zero area. Although not admitting such a concrete representation, the "ideal boundary" of a Riemann surface of infinite genus has begun to play a somewhat similar role in the classification of open Riemann surfaces and in the study of existence problems on them, and we propose to outline here a method for representing this ideal boundary.

Our starting point will be the class BD of those piecewise smooth functions f defined on W which are bounded and have a finite Dirichlet integral $D[f]$. Since the product of two functions of the class BD is again a function of this class, we see that these functions form a commutative ring with unity. We introduce a topology into this ring by saying that a sequence f_i converges to f in BD if $|f_i|$ is uniformly bounded and f_i converges to f uniformly on every compact subset of W while $D[f - f_i]$ tends to zero. A set A is then said to be closed in this topology if $f_i \rightarrow f$ and $f_i \in A$ imply $f \in A$. (Note that this topology does not satisfy the first countability axiom.)

The set K consisting of all those BD functions which vanish outside a compact set forms an ideal in the ring BD. If we denote by $\overline{K}$ those functions which are limits in BD of sequences from K, then $\overline{K}$ is not only closed in our topology but also weakly closed [1].

If the function 1 belongs to $\overline{K}$ we say that W is parabolic, and otherwise that it is hyperbolic. This condition is equivalent to the usual conditions concerning the non-existence of a Green's function or the validity of the unrestricted maximum principle. In the case of a parabolic surface $\overline{K} = \text{BD}$, while for a hyperbolic surface we have for each $f \in \text{BD}$ the unique decomposition

$$f = f_K + u$$

where $f_K \in \overline{K}$ and $u \in \text{HBD}$, that is, a harmonic function of the class BD [1]. Since it can be proved that the projection

$$\pi : f \rightarrow u$$

is continuous in the BD topology, we may write

$$BD = \overline{K} + HBD \quad .$$

We now abandon the BD topology and introduce a stronger topology (more open sets) given by the norm

$$||f|| = \sup |f| + D[f].$$

The projection π is norm decreasing in this norm, and so must remain continuous if we complete the space BD under this norm. Since HBD is already complete in this norm, the completion has merely added more functions to the kernel of π which we again denote by $\overline{K}$.

This completed space is now a Banach algebra A and we may, according to the Gelfand theory, represent this as a ring of continuous functions on the compact space W^* consisting of the regular maximal ideals of A. Since those maximal ideals of A which do not contain K are in one-to-one correspondence with the regular maximal ideals of K, it follows that the set of these ideals is homeomorphic to W itself. On the other hand, the set Γ of maximal ideals of A which contain K form a closed non-dense subset of W^* which we regard is the ideal boundary of W. The set Δ of maximal ideals containing $\overline{K}$ is a closed subset of Γ which we call the harmonic boundary of W. If we define a multiplicative structure in HBD by means of the projection π, the maximal ideals of HBD are homeomorphic to Δ, and HBD is isomorphic to a ring of continuous functions on Γ.

The relationship of the harmonic function on W to functions on W^* becomes clearer if we consider the class HBD^* of those harmonic functions which are uniform limits of HBD functions. For then the ring HBD^* is isomorphic to the ring of all continuous functions on Δ with the uniform norm, and the Dirichlet problem has the following formulation:

Given a continuous function f defined on Δ, there exists a unique function u which is defined and continuous on W^*, equal to f on Δ and harmonic in W. On W the function u belongs to the class HBD^*.

It is clear from the nature of the ideals of A that a separation of W into disjoint parts by a compact curve also separates Γ. Hence the part of Γ added to an end of the surface depends only on the end and not on the rest of the surface.

If we take W to be the unit circle $|z| < 1$, then Δ consists of the circle $|z| = 1$ together with points corresponding to various modes of tangential approach to $|z| = 1$. If we wish to obtain the usual boundary of the unit circle, we must consider instead of A the closed subalgebra A_1 generated by the functions x and y. Just what generators should be chosen in the general case is an open question, but I believe it would be most useful to take solutions of certain extremal problems corresponding to the classical slit mappings. This gives the usual compactification in the case of finite Riemann surfaces.

In the case of a parabolic surface it seems clear that the proper generators to take are those functions which are harmonic outside compact sets. Using the subalgebra generated by them, one obtains the usual boundary for a planar parabolic surface.

By considering a hyperbolic surface with no HD functions on it, we see that on the other hand the harmonic boundary Δ of a Riemann surface can consist of a single point.

It is an open question whether or not the compactification of a single topological end can contain both points of $\Gamma - \Delta$ and Δ, i.e., whether or not a hyperbolic end can have points of $\Gamma - \Delta$ in its compactification.

BIBLIOGRAPHY

[1] HEINS, M., "Riemann Surface of Infinite Genus." To appear in Annals of Math. 55 (1952) pp. 296-317.

[2] ROYDEN, H. L., "Harmonic functions on an open Riemann Surface," to appear in Transactions Amer. Math. Soc. 73 (1952) pp. 40-94.

TOPOLOGICAL METHODS ON RIEMANN SURFACES
PSEUDOHARMONIC FUNCTIONS

James A. Jenkins and Marston Morse

§1. INTRODUCTION

The topological analysis which has in recent years been associated with extensions of the theory of functions of a complex variable and Riemann surfaces includes among other aspects the following divisions:

I. The point set characterization of an interior transformation by Stoïlow [13] and Whyburn [15]. This has been followed very recently by a similar characterization by Y. Tôki [14] of a pseudoharmonic function. A characterization by the same author of a pseudoconjugate seems questionable.

II. The structural analysis of the relations between the zeros, poles and branch point antecedents of an interior transformation under topologically defined boundary conditions including boundary conditions definable in terms of locally simple images of a Jordan curve, and an antecedent theory involving the logarithmic poles and critical points of a pseudoharmonic function. See [10], [11]. Extensions on a general Riemann surface remain to be made.

III. A theory of the deformation classes of single-valued meromorphic functions defined in an arbitrary simply connected plane region R. Here the zeros, poles, and branch point antecedents are supposed fixed during the deformation, as well as R. The theory extends to interior transformations. There is no difference between the purely topological theory for interior transformations and the theory for meromorphic transformations in so far as the numerical characteristics of a deformation class are concerned, but an essential difference arises when one takes account of the total covering of the Riemann sphere by a set of functions containing one model for each deformation class. Extensions exist when the zeros, poles, and branch point antecedents are deformed. See [4], [12].

IV.a The existence of a pseudoharmonic function U on an open Riemann surface when trajectories having the characteristic top (topological) properties of level lines of a harmonic function are proposed as level sets of U. Something of the history of this problem will be discussed in §2. This is the problem which is attacked in this paper in somewhat more general form than previously. The history of the problem of the existence of a pseudoconjugate to U will also be discussed in §2, and a detailed top proof of the existence of a pseudoconjugate to U given in a later paper [6] under the appropriate conditions.

IV.b There remains the problem of the existence and top characterization of pseudoharmonic functions and their pseudoconjugates in the compact case including the case where simple boundary conditions are given. In [5] the present authors have given a complete solution of these problems for the case in which the functions are defined on a closed circular disc, and the boundary values of U have a finite number of extrema. The crucial net N of level lines meeting the critical points of U admits a very simple model in the hyperbolic plane. With Poincaré we use the unit disc D as a model for the hyperbolic plane. The essential condition on N is that the closure in $\overline{D}$ of no subset of lines in N is bounding in $\overline{D}$.

DEFINITIONS. Let G^* be an arbitrary open Riemann surface. Let U_0 be harmonic on G^* in a neighborhood H of $q \in G^*$ and not identically constant. Let ψ be a top mapping of a neighborhood H_0 of q onto H. Then $U_0\psi$ is termed PH (pseudoharmonic) at q. A function U is termed PH in a region $R \subset G^*$ if U is PH at each point of R.

Let φ be analytic in a neighborhood H of $q \in G^*$ and not identically constant and let ψ be a sense-preserving top mapping of a neighborhood H_0 of q onto H. Then $\varphi\psi$ is termed interior at q. We term f interior over a region $R \subset G^*$ if f is interior at each point of R.

If U and V are PH over a region $R \subset G^*$ and if $U + iV$ is interior over R then V is termed a pseudoconjugate of U over R.

§2. HYPOTHESES AND HISTORY

Let G^* be an oriented open connected Riemann surface metricized in any convenient way. Let ω be a discrete set of points in G^* and set $G^* - \omega = G$. Let F be a family of elements α, β, γ etc., open arcs or top circles in G which include one and only one $\alpha \in F$ meeting each point $p \in G$. In a sense to be more explicitly defined in §3 we suppose that F has the local top properties on G^* of the level sets of a function U PH over G^* with ω the set of critical points of U. [10, §3].

The non-recurrence hypothesis. In this report an open arc in G is the 1-1 continuous image in G of an open interval. Kaplan [7] has studied families F in the z-plane E restricting open arcs to homeomorphic images in E of an open interval. Such restricted arcs are never their own limit sets in the sense of dynamical systems. They are <u>non-recurrent</u> as we shall say. Admitting no open α which is recurrent Kaplan showed that when $G^* = E$ and $\omega = 0$ the family F gives the level sets of a PH function.

In the z-plane E the connected open level sets of a single-valued PH function are obviously never self-limiting. But to make this hypothesis of non-recurrence would exclude harmonic functions of frequent occurrence. In particular if u_1 and u_2 are independent elliptic integrals of the first kind, then for almost all choices of a real constant c, the level curves of $\mathcal{R}(u_1 + c\, u_2)$ are their own limit sets. When $G^* = E$ we <u>prove</u> that no $\alpha \in F$ is its own limit set nor a top circle. Kaplan's proof that no $\alpha \in F|E$ is a top circle depends strongly on the assumption that no $\alpha \in F|E$ is its own limiting set. [7,I, p. 162].

I. Making no assumption of non-recurrence it can be shown, for a general ω and under general boundary conditions which include the special case $G^* = E$, that there exists a function U, PH over G^* with each $\alpha \in F$ a level set of U and with a critical point at each point of ω.

Boothby [1] establishes this theorem less generally under the hypothesis of non-recurrence and in the special case $G^* = E$. Our proofs were constructed before Boothby's were available and are of altogether different character. We do not reduce the case of a general set ω to the case $\omega = 0$, as does Boothby but treat the general ω at once.

Problem II concerns the existence of a function v PH over G^* and pseudoconjugate to a u given as PH over G^*. Kaplan affirmed in 1941, [9], that in case $\omega = 0$ the family $F|E$ is homeomorphic with the family of trajectories of a suitably chosen differential system.

$$\frac{dx}{dt} = p(x, y) \qquad \frac{dy}{dt} = q(x, y)$$

where p and q are of class C' over E. The method of proof suggested does not seem sufficiently explicit to be final, nor does the method suggested later in [8]. The present report will be followed by a detailed proof of the existence of v when $G^* = E$ and ω is general. Our method is topological and depends upon properties of μ-length of curves in an abstract metric space, particularly in situations where the curves are not simple, [10, §27]. Boothby's proof [2] of the existence of v when $G^* = E$ depends upon the above affirmation of Kaplan.

The methods of this report include the introduction of an F-vector distribution over G in the case of a general open Riemann surface. These F-vectors are simple subarcs of elements of F and vary continuously in the sense of Fréchet with their initial points on a two fold covering M of G^* When $G^* = E$ the F-vector distribution reduces to two distributions, distinct, single-valued, and continuous over $G = G^* - \omega$ with null limits at points of ω. When $G^* = E$ these F-vectors may be replaced by ordinary vectors joining their end points. The special indices of Hamburger are not needed since one can apply the methods of Poincaré and Bendixson to our F-vectors.

§3. THE FAMILY F

An arc, open arc or top circle in G^* is the 1-1 continuous image on G^* of a closed interval, open interval, or circle, respectively. These elements are to be distinguished from parameterized curves, which are mappings, - not sets, and curves, which are equivalence classes of such mappings. These distinctions become essential when the curves are not simple.

DEFINITION OF F. F shall be an ensemble of elements α, β, γ, etc. which are either open arcs or top circles in $G = G^* - \omega$, such that there is one and only one element in F meeting each point $p \in G$ and such that F has the following local property. Let D be the open disc $(|w| < 1)$ in the complex w-plane. Let $\overline{A}$ denote the closure relative to G^* of a set $A \subset G^*$. With each point $p \in G^*$ there shall be associated an F-<u>neighborhood</u> X_p of p with $\overline{X}_p \subset G \cup p$, and a homeomorphic mapping T_p of $\overline{X}_p$ onto $\overline{D}$ under which p goes into $w = 0$ and the maximal open arcs of $F|X_p$ go into the maximal open level arcs of $\mathcal{R} w^n$, $n > 0$, in D (with $w = 0$ excluded when $n > 1$.) We term n the <u>exponent</u> of p. The exponent of p shall be 1 for $p \in G$ and > 1 for $p \in \omega$. Points $p \in \omega$ are termed singular points of F. We term $w \in D$ the <u>canonical parameter</u> of its antecedent in X_p.

The open arcs of $F|(x_p - p)$ which have limiting end points at $p \in G^*$ are termed F-<u>rays</u> of X_p incident with p. They are $2n$ in number thereby showing that n is independent of the choice of X_p. These F-rays of X_p divide $X_p - p$ into $2n$ open regions termed F-<u>sectors incident with</u> p.

<u>Right</u> N. With each $p \in G$ one can also associate a neighborhood N_p of p with $\overline{N}_p \subset G$ and a homeomorphic sense preserving mapping of $\overline{N}_p$ onto a square $K : (-1 \leq u \leq 1)(-1 \leq v \leq 1)$ such that p goes into the origin in K and the maximal subarcs of $F|\overline{N}_p$ go into arcs $u = c$, $-1 \leq v \leq 1$ where the constant c ranges over the interval $[-1, 1]$. We refer to N_p as a <u>right</u> N of p and term u and v <u>canonical coordinates</u> of the antecedent in N_p of (u, v) in K.

Transversals. By a transversal λ is meant an open arc in G whose intersection with any right N with canonical coordinates (u, v) locally has the form $v = \varphi(u)$ where φ is single-valued and continuous. By the principal transversal of a right N is meant the open arc in N on which $v = 0$, $-1 < u < 1$.

F-vectors. Any sensed subarc A of an $\alpha \in F$ will be called an F-vector. By definition an F-vector is simple, closed, and never a top circle. F-vectors will be represented by capital letters A, B, V, etc.

LEMMA 3.1. Each F-vector A is in some right N.

Let B be an F-vector which contains A in its interior. Let B be given as the 1-1 continuous image of an interval $a \leq t \leq b$, with B(t) the image in B of t in [a, b]. There clearly exists a sequence $a = t_0 < t_1 < \dots < t_n = b$ of values of t defining points $p_r = B(t_r)$ $(r = 0,\dots,n)$ in B such that p_i and $p_{i-1} (i=1,\dots,n)$ lie in a common right N_i. Let transversals $\lambda_0, \lambda_1, \dots, \lambda_n$ be successively chosen so that λ_r meets $p_r (r = 0,\dots,n)$, λ_{i-1} and λ_i are in $N_i (i = 1,\dots,n)$, and $\lambda_r \cap \lambda_i = 0$ for $r \neq i$. Let $\lambda_0(s)$ be the point on λ_0 with canonical coordinate $s = u$ in N_1 and with $\lambda_0(0) = p_0$. Taking i successively equal to 1, 2, ..., n let $\lambda_i(s)$ be the point (if any exists) on λ_i on a subarc A_i^s of an element of F such that A_i^s joins $\lambda_{i-1}(s)$ to $\lambda_i(s)$ in N_i. If $e > 0$ is sufficiently small and $-e \leq s \leq e$, $\lambda_r(s)$ is well defined and continuous for $s \in [-e, e]$ and $r = 0,\dots,n$, since λ_i and λ_{i-1} lie in the same right N_i. Since A is simple, for e sufficiently small

$$A_r^s \cap A_i^{s'} = 0 \quad (s, s' \in [-e, e]) \quad (r < i)$$

except that $A_r^s \cap A_{r+1}^s = \lambda_r(s)$. We suppose e so chosen.

Since λ_i and λ_{i-1} lie in the same right N_i the subset of N_i covered by the F-vectors $A_i^s | s \in [-e, e]$ is the homeomorph of a rectangle

$$K_i : [-e \leq x \leq e] \quad [i-1 \leq y \leq i] \quad [i = 1.,,,n]$$

with the arc A_i^s mapped onto the arc $x = s$, $i-1 \leq y \leq i$, of K_i for each $s \in [-e, e]$ in such a manner that $\lambda_{i-1}(s)$ and $\lambda_i(s)$ go respectively into the points $(x, y) = (s, i-1)$ and (s, i) of K_i. The union of the rectangles K_i is a rectangle

$$H : (-e \leq x \leq e) \quad (0 \leq y \leq n)$$

which can be mapped linearly onto the square $K : (-1 \leq u \leq 1)(-1 \leq v \leq 1)$ in such a manner that the arcs $x = s$ in H (s constant) go into the arcs $u = c$ of K $(-1 \leq v \leq 1)$. The union of the arcs $A_i^s | s \in [-e, e]$, $i = 1,\ldots,n$, is thereby mapped homeomorphically onto K and becomes the closure of a right N which contains A as a subarc of the arc $u = 0$ in N.

§4. THE F-VECTOR FIELD (V)

The Fréchet distance between two sensed arcs in an arbitrary metric space is well defined. Let the space of F-vectors in G^* be metricized by Fréchet distance.

The limit $\delta(p)$. For $p \in G$ let $\alpha_p \in F$ meet p. If α_p is an open arc, p divides α_p into two half open arcs π' and π'' with p as initial point. The μ-lengths [10, §27] $\mu(\pi')$ and $\mu(\pi'')$ of π' and π'' are well defined. Let

$$\delta(p) = \min [\mu(\pi'), \mu(\pi'')] .$$

In case α_p is a top circle let A' and A'' be F-vectors on α_p with initial point p, with equal μ-lengths t such that A' and A'' intersect only in p. Let $\delta(p)$ be the sup of such values of t.

LEMMA 4.1. The function δ is lower semi-continuous over G.

Given $p_0 \in G$ let δ_0 be a constant such that $0 < \delta_0 < \delta(p_0)$. Let B' and B'' be the two F-vectors on α_{p_0} with initial point p_0 and μ-length δ_0. Let $|B'|$ and $|B''|$ be the arcs carrying B' and B'' respectively. Then $|B'| \cup |B''|$ is a simple arc. By Lemma 3.1 this arc lies in a right N. In a sufficiently small neighborhood of p_0 in N it is clear that $\delta(p) > \delta_0$ and the lemma follows.

The modul $\sigma(p)$. Since $\delta(p) > 0$ at each point of G, and δ is lower semi-continuous it follows from the proof of Satz I [3, pp. 159-162] that there exists a function σ with value $\sigma(p)$ continuous and bounded over G, and such that $0 < \sigma(p) < \delta(p)$. We can also suppose that $\sigma(p) \to 0$ as p tends to any singular point q of F.

To see this let an F-neighborhood X_p be associated with each singular point q of F in such a manner that for any two singular points q and r of F, $X_q \cap X_r = 0$. Let $w|$ $(|w| < 1)$ be the canonical complex parameter representing $p \in X_q$, with q represented by $w = 0$. Let $f(w)$ be the value of $\sigma(p)$ when w represents p. We then replace $f(w)$ by $|w|f(w)|(|w| < 1)$. If such an alteration of $\sigma(p)$ is made in each X_q the resultant values $\sigma(p)$ will tend to zero as p tends to any singular point q of F. The new function σ will be continuous with $0 < \sigma(p) < \delta(p)$.

<u>The F-vector field</u> (V). With each point $p \in G$ and $\alpha_p \in F$ meeting p we associate the two F-vectors V'_p and V''_p on α_p with initial point p and μ-length $\sigma(p)$. The totality of such special F-vectors V is called the F-vector field (V). The field (V) will be represented in a 1-1 way by an open 2-fold covering M of G.

<u>The covering manifold</u> M. The points of M shall be the pairs (p, V_p), where p is in G and V_p is V'_p or V''_p. A neighborhood of $(q, V_q) \in M$ shall be any set of pairs (p, V_p) such that for some right N_q of q, p ranges over a neighborhood of q in N_q, the terminal point of V_p varies continuously with p in N_q and $|V_p|$ is in N_q. It is clear that M is a Hausdorff space.

The point $p \in G$ is termed the <u>projection</u> of (p, V_p) into G and (p, V_p) is termed a <u>projection antecedent</u> in M of $p \in G$. A sufficiently small neighborhood H of a point $p_0 \in G$ has two disjoint projection antecedents in M each the homeomorph of H under projection. Corresponding to any arc β in G there exist two unique disjoint arcs β' and β'' in M which project homeomorphically onto β.

LEMMA 4.1. The manifold M is either connected or consists of two disjoint connected manifolds $M^{(1)}$ and $M^{(2)}$ each of which projects homeomorphically onto G.

Let (p, V'_p) and (p, V''_p) be the two points in M which project into $p \in G$. Let (q, V_q) be an arbitrary point in M with $p \neq q$. Let β be a sensed arc in G which joins p to q, and let β' and β'' be the disjoint sensed arcs in M which project homeomorphically onto β. Then the terminal point of β' or else of β'' must be (q, V_q) so that every point of M is arcwise connected to (p, V'_p) or else to (p, V''_p). If M is not connected a component $M^{(1)}$ of M contains but one projection antecedent of a given point of G, so that the projection of $M^{(1)}$ onto G is 1-1. On $M^{(1)}$ both projection and its inverse are open transformations so that projection is a homeomorphism on $M^{(1)}$.

COROLLARY 4.1. If M is disconnected there are two single-valued mappings from G to the field (V) with values $V_p^{(1)}$ and $V_p^{(2)}$ at $p \in G$ such that $V_p^{(i)}$, $i = 1, 2$, varies continuously in the sense of Fréchet with $p \in G$.

A disconnected M is the union of two disjoint connected components with points $(p, V_p^{(1)})$ and $(p, V_p^{(2)})$ respectively. That $V^{(1)}$ and $V^{(2)}$ are continuous over G follows from the definition of neighborhoods on M.

<u>Indicatrices in</u> G. Let a top circle C on G which is bounding on G be arbitrarily sensed. Recall that G^* and hence G is orientable. Any top circle C' on G, bounding on G, can be isotopically deformed on G into C, and will be given the <u>unique</u> sense derived from C under this deformation. Such sensed top circles will be termed indicatrices in G.

<u>F-vector indicatrices</u>. Let B be an arbitrary F-vector on G, lying in a right N. Let h be an inner arc of B sensed as in B. Let C be a top triangle which is an indicatrix of G in which h is one "side" and the other two "sides" are transversals in N. We term C a B-indicatrix. If B' and B" are F-vectors similarly sensed in a right N, any B'-indicatrix can be isotopically deformed (always understood with preservation of sense) in N into any B"-indicatrix through B-indicatrices such that B is an F-vector in N. More generally if an F-vector B' is isotopically deformable through F-vectors (always understood with preservation of sense) into an F-vector B" then any B'-indicatrix is isotopically deformable through F-vector indicatrices into any B"-indicatrix. For the desired deformation can be effected as a resultant of a finite set of deformations each made in a right N.

THEOREM 4.1. If the elements of F are level sets of a single-valued PH function U then the manifold M is disconnected.

Given $V \in (V)$ let C be a V-indicatrix. Then

$$U(p)|(p \in V) - U(q)|(q \in C-V) \tag{4.1}$$

is of one sign independent of $q \in C-V$. This sign is clearly invariant under any isotopic deformation of C as an indicatrix of some F-vector. If V is isotopically deformed through F-vectors in (V) so as to return to a vector V' with initial point in common with V then $V = V'$. For, if C' is a V'-indicatrix, the sign in (4.1) will be the sign of

$$U(p)|(p \in V') - U(q)|(q \in C'-V') \tag{4.2}$$

Now $|V| \cup |V'|$ is an arc in a right N on an element $\alpha \in F$. If $V \neq V'$, C and C' would intersect different components of $N - \alpha$ so that (4.1) and (4.2) could not have the same sign. Hence $V = V'$ and it follows that M has two disjoint components.

LEMMA 4.2. If X_p is a sufficiently small F-neighborhood of a singular point $p \in G^*$, then the projection antecedent of $X_p - p$ in M is the union of two disjoint open sets in M each of which projects homeomorphically into $X_p - p$.

Let Y_p be an arbitrary F-neighborhood of p such that $F|Y_p$ is the top image of the set F_w of open level arcs of $\mathcal{R}w^n|(0<|w|<1)$. Let $e>0$ be so small that the subset X_p of Y_p on which $|w|<e$ is an F-neighborhood of p such that each F-vector $V \in (V)$ with initial point in $X_p - p$ is in Y_p. The function $\mathcal{R}w^n$ is PH; it accordingly follows from the properties of F-vector indicatrices (as exhibited in the proof of Theorem 4.1) that any isotopic deformation of a subarc β of an element in F_w which is through such subarcs and which returns β to an arc with the same initial point and μ-length, returns β to itself. Hence the projection antecedent of $X_p - p$ has two disjoint components. The projection of each component into M is then 1-1, open, and has an open inverse, and hence is a homeomorphism.

THEOREM 4.2. If R is any simply-connected region of G^* then the projection antecedent of $R - \omega$ in M is the union of two disjoint open sets each of which projects homeomorphically into $R - \omega$.

Lemma 4.2 is essential for the proof of Theorem 4.2 but it must be supplemented by the statement that the inverse projection of a sufficiently restricted F-neighborhood X_p of a $p \in G$ is the union of two disjoint open sets in M each of which projects homeomorphically into M. Theorem 4.2 is a consequence of the well known monodromy law for simply-connected regions as applied to the two-valued inverse projection from $R - \omega$ to M. The fact that $R - \omega$ is not simply-connected if $R \cap \omega \neq 0$ is immaterial, since the inverse projection of G onto M has two distinct, single-valued, continuous branches over suitable punctured neighborhoods $X_p - p$ of each point $p \in R$.

§5. COHERENT SENSING

Let each $\alpha \in F$ be given a sense. The resulting family F^s of sensed open arcs and top circles will be called a <u>sensed image</u> of F. We shall refer to a continuous deformation Δ of an F-vector A in the Fréchet space of F-vectors. We shall understand that each image of A under Δ is <u>sensed by</u> Δ, that is, that the sense of the image shall be determined by the images under Δ of the initial and final points of A. We say that F^s is <u>coherently</u> sensed if any continuous deformation Δ of an F-vector A <u>initially</u> sensed by F^s, through F-vectors sensed by Δ, is necessarily a deformation through F-vectors sensed by F^s.

THEOREM 5.1. If M is disconnected there exist two distinct coherently sensed images F^s of F.

Let $V_p \in (V)$ be an F-vector, with initial point p, in one of the two single-valued continuous distributions of F-vectors of (V) over G possible when M is disconnected (Cor. 4.1). Let F^s be defined by assigning to the $\alpha \in F$ meeting a $p \in G$ the sense of V_p. We shall show that F^s is coherently sensed.

To this end let Δ be a continuous deformation of an F-vector A through F-vectors A^t, $0 \leq t \leq 1$, where $A = A^0$ is initially sensed by F^s. Let $p(t)$ be the initial point of A^t. Let $V(A^t)$ be the F-vector in (V) with the initial point $p(t)$ and sense of A^t. Then $V(A^t) = V_{p(t)}$ when $t = 0$; on increasing t continuously this relation is seen to hold for each $t \in [0, 1]$. Hence A^t has the sense of $V_{p(t)}$, that is the sense of F^s. Hence F^s is coherently sensed.

Theorem 5.1 taken with Theorem 4.2 gives the following.

COROLLARY 5.1. If G^* is simply-connected there exist two distinct coherently sensed images F^s of F.

COROLLARY 5.2. Suppose M disconnected and let λ be a transversal of a right N. No subarc β of an $\alpha \in F$ which is not in N and does not cross λ can have its end points in the same component of $N - \lambda$.

Since M is disconnected F has a coherently sensed image F^s. (Th. 5.1). Let β take its sense from F^s. Terminal subarcs of β in N, if sensed as in β, and if in the same component of $N - \lambda$ are oppositely sensed in terms of the canonical coordinates of N. Either sensed terminal subarc of β can be isotopically deformed into the other reversed in sense through F-vectors in N. Hence F^s cannot be coherently sensed. From this contradiction we infer the truth of Corollary 5.2.

§6. INDICES $G^* \subset E$

In this section we suppose G^* in the z-plane E and that F has a coherently sensed image F^s. Let $(V)_s$ be the subset of F-vectors in the field (V) with senses derived from F^s. Each F-vector A determines a chord in the z-plane joining the initial to the final point of A. Let <u>arc</u> A stand for the multiply-valued angle which this chord makes with the positive x-axis $(z = x + iy)$.

<u>The</u> F-<u>index</u> $I(C)$. Let C be a top circle in G. Let $V(p)$ denote the F-vector in $(V)_s$ with initial point $p \in G$. As p traces C in its counter-clockwise sense a continuous branch of arc $V(p)$ has an

algebraic increment $2m\pi$. We then assign C an F-index $I(C) = m$. If the coherently sensed image F^s of F'' is replaced by the oppositely sensed image of F the index $I(C)$ is unchanged. For this reason we term $I(C)$ the F-index or V-index of C.

A boundary index. Suppose that R is a connected region in G^* bounded by r non-intersecting top circles $C_1,\dots,C_r$ in G, with $C_2,\dots,C_r$ interior to C_1 in E. Let βR denote the boundary of R. We introduce the boundary index $I(\beta R) = I(C_1) - I(C_2) \quad - I(C_n)$.

The F-index $I(z_0) \mid (z_0 \in G^*)$. Let C be any top circle in G whose interior contains z_0 and, except at most for z_0, is in G. Set $I(z_0) = I(C)$, noting that $I(C)$ is independent of the choice of C conditioned as in the preceding sentence; for any two admissible choices C' and C'' of C can be isotopically deformed into each other through admissible choice of C. We shall prove the following.

(a) *If* $F \mid (0 < |z| < c)$ *is given by the level arcs of* $\mathcal{R} z^n \mid (0 < |z| < c)$ $(n > 1)$ *then* $I(z_0) = 1 - n$ *when* $z_0 = 0$.

Let $V(\theta)$ be the F-vector in $(V)_s$ at the point $z(\theta) = be^{i\theta}/2$, $0 < \theta < 2\pi$, where $0 < b < c$ and b is so small that $|z| < c$ on $V(\theta)$. Let $Z(\theta)$ be the unit vector positively tangent to the $\alpha \in F^s$ meeting $z(\theta)$. It is possible to choose continuous branches of arc $V(\theta)$ and arc $Z(\theta)$ such that $|\text{arc } V(\theta) - \text{arc } Z(\theta)| < \text{const} < \pi$ $(0 \leqq \theta < 2\pi)$. It follows that $2\pi I(z_0)$ equals the algebraic variation of arc $Z(\theta)$ as θ increases from 0 to 2π. This variation is $1-n$ so that (a) is true.

Note that $I(z_0)$ in (a) is independent of the choice of (V) among admissible (V).

LEMMA 6.1. *If* z_0 *is a point of* G *of exponent* n *then* $I(z_0) = 1-n$.

Suppose $z_0 = 0$. With z_0 there is given a sense preserving homeomorphic mapping T of the closure of an F-neighborhood X of z_0 onto a disc $\bar{D}$ $(|z| \leqq 1)$ leaving $z_0 = 0$ fixed, and such that when $n > 1$, $F \mid (X - z_0)$ is given by the open level arcs of $\mathcal{R}(T(z))^n \mid (0 < |T(z)| < 1)$ and when $n = 1$ by the open level arcs of $\mathcal{R} T(z) \mid (|T(z)| < 1)$.

One can deform T into the identity through a family of homeomorphisms T_t, $0 \leqq t \leqq 1$ each of which is a mapping of $\bar{X}$ into E leaving z_0 fixed. To this end let T^* be a homeomorphic mapping of $\bar{X}$ onto $\bar{D}$ such that T^* is directly conformal over X. By virtue of a theorem of Tietze one can deform T into T^* through a family of homeomorphic mappings of $\bar{X}$ onto $\bar{D}$. One can also deform $T^* \mid \bar{X}$ into the identity using a family of homeomorphic mappings of $\bar{X}$ into E which are conformal on X, leave z_0 fixed, and are such that $T^* \beta X$ is isotopically deformed onto βX. Thus T is deformed into the identity.

Let $V(z)$ be the F-vector of $(V)_s$ with initial point $z \in G$. There exists a neighborhood $U \subset X$ of z_0 such that for $z \in U - z_0$, $V(z)$ is in X. For $z \in U - z_0$ and each t set $w = T_t(z)$ and $T_t\, V(z) = V_t(w)$. Here $w_0 = z_0 = 0$. Let $I_t(z_0)$ be the V_t-index of z_0. It is clear that $I_t(z_0)$ is independent of t for $0 \leq t \leq 1$, that $I_1(z_0) = I(z_0)$, and when $n > 1$ $I_0(z_0) = 1 - n$ by virtue of (a). Hence $I(z_0) = 1 - n$ when $n > 1$.

That $I(z_0) = 0$ when $n = 1$ follows on evaluating $I(z_0)$ as an index $I(C)$ of a circle $C \mid (|z| = e)$ for which e is sufficiently small.

The classical methods of treating vector fields are adequate to prove the following.

THEOREM 6.1. If R is an open region of $G^* \subset E$ bounded by a finite number of non-intersecting top circles each in G, and if F admits a sense coherent image the F-index of βR equals the sum of the indices of the singular points of F in R.

COROLLARY 6.1. It is impossible for a top circle C in F to bound in $G^* \subset E$.

Suppose C bounded a Jordan region $H \subset G^*$. Then $\overline{H}$ would be interior to a Jordan region $R \subset G^*$ and $F \mid (R \cap G)$ would admit a sense coherent image. One could then apply Theorem 6.1 assuming that $G^* = R$. The index $I(C)$ is merely the angular order [10, §18] of C and so equals 1.

By Theorem 6.1 $I(C)$ would equal the sum of the indices of the singular points of F in H. This sum is negative or zero and so cannot equal $I(C) = 1$. We infer the truth of Corollary 6.1.

§7. NON-RECURRENCE WHEN $G^* = E$

In this section we assume that G^* is the finite z-plane E.

LEMMA 7.1. If $G^* = E$ no F-vector A intersects a transversal λ of a right N in more than one point.

(a) It will be sufficient to show that there exists no F-vector A which intersects λ solely in the end points of A.

We can suppose that λ is the principal transversal of a right N so that, in terms of the canonical coordinates (u, v) of N, $v = 0$ and $-1 < u < 1$ on λ. Since $G^* = E$ it follows from Theorem 4.2 that the covering manifold M of §4 is disconnected, and then from Corollary 5.2 that the intersection of A with $\overline{N}$ (if A exists contrary to (a)) is the union of two disjoint F-vectors on which $v \geq 0$ and $v \leq 0$ respectively.

In terms of the canonical coordinates (u, v) in N let $(a, 0)$ and $(b, 0)$ be the initial and final points of A in λ. For definiteness suppose that $v \geq 0$ on the initial subarc of A in $\overline{N}$. Set $B = A - (A \cap N)$. Then B is an F-vector with end points $(a, 1)$ and $(b, -1)$ on βN. We shall show how to modify $F|N$, leaving elements $\alpha \in F$ otherwise unchanged, in such a manner that the existence of B implies the existence of a top circle in the modified F.

Let T be a homeomorphic mapping of $\overline{N}$ onto itself defined as follows. The edges $u = \pm 1$ of N shall be pointwise invariant. The edges $v = \pm 1$ shall be so mapped onto themselves that the points $(0, 1)$ and $(0, -1)$ go respectively into $(a, 1)$ and $(b, -1)$. Let $T|N$ be further defined so that an arc $u = c$, $-1 \leq v \leq 1$, in N is carried linearly into the straight arc which joins the T-images of its end points. Let F' replace F, taking $F|(G - N) = F'|(G - N)$ and $F'|N = TF|N$. In F' the end points of B will be connected by a straight arc in N. The resulting top circle in F' is impossible by virtue of Corollary 6.1. Hence statement (a) is true and the lemma follows.

LEMMA 7.2. *If $G^* = E$ there is a first and a last point in which any sensed $\alpha \in F$ meets a given compact subset G_0 of G, whenever α meets G_0.*

There is a finite set of right $N\ N_i$, $i = 1,2,\ldots,n$ in whose union R the set G_0 lies. If α meets N_i it follows from Lemma 7.1 that there is a first point p_i and a last point q_i in which α meets $\overline{N}_i$. In the ensemble of points p_i and q_i $(i = 1,\ldots,n)$ on α there is a first and a last point p and q on α. The points p and q on α are the end points of an F-vector on α which contains $\alpha \cap G_0$. Since neither p nor q is in G_0 the lemma follows.

<u>The family</u> F^*. We shall define F^* for a general open Riemann surface G^*. F^* shall consist of elements h, k, m, etc., which are top circles or open arcs in G^* with the following properties. If a non-singular point p is in h, h shall contain the $\alpha_p \in F$ meeting p, and any limiting end point or end points of α_p in G^*. If a singular point q is in h, h shall contain just two of the $\alpha \in F$ which have q as a limiting end point. An $h \in F^*$ which is also in F will be called <u>non-singular</u>.

A top circle $h \in F^*$ is compact and so contains at most a finite set of singular points of F and hence at most a finite set of $\alpha \in F$.

<u>The graph of</u> F^*. The union of ω and of the $\alpha \in F$ with at least one limiting end point at a singular point of F is called the graph of F^*. Let H be a component of the graph of F^*. Let the $\alpha \in F \mid H$ be enumerated, α_i, $i = 1,2,\ldots,$ and let each top circle h in H be

assigned an <u>order</u> equal to the maximum index i of the α in $F \mid h$. Let a top circle of minimum order (if any exists) be deleted from H to form a graph H_1. From H_1 let a top circle of minimum order (if any exists) be deleted. No singular point q shall be deleted as long as there remain $\alpha \in F$ incident with q. Continuing this process indefinitely one either exhausts H or obtains a subgraph H^* which contains no top circles. If H^* is not empty any $h \in F^*$ in H^* is formed from a sequence of $\alpha \in F$ of one of the forms.

(7.1) $$\ldots \alpha_{-2}\, \alpha_{-1}\, \alpha_0\, \alpha_1\, \alpha_2 \ldots$$

(7.2) $$\ldots \alpha_{-2}\, \alpha_{-1}\, \alpha_0; \quad \alpha_0\, \alpha_1\, \alpha_2 \ldots$$

(7.3) $$\alpha_0\, \alpha_1 \ldots \alpha_n$$

The terminal α_0 or α_n must, in one sense, have no limiting end point in G^*. One sees that every $\alpha \in F$ is an $h \in F^*$ or is in an $h \in F^*$. A given $\alpha \in F$ may be in many $h \in F^*$.

<u>Transversal</u> <u>rays</u>. Let X_p be an F-neighborhood of a singular point p such that $F \mid (X_p - p)$ is given as the top image of the set of level arcs of $\mathcal{R} w^n$ $(0 < |w| < 1)$. Each of the $2n$ open arcs

$$(0 < |w| < 1) \quad (\text{arc } w = \frac{r\pi}{n}) \quad (r = 0,\ldots,2n-1)$$

is a transversal of $F \mid X_p$ and will be called a transversal ray incident with p. When $G^* = E$ it follows from Lemma 7.1 that no $\alpha \in F$ can intersect a transversal ray of $F \mid X_p$ more than once. Hence a non-singular $h \in F^*$ is bounded from any given singular point p when $G^* = E$.

LEMMA 7.3. If $G^* = E$ there is a first and a last point in which any sensed non-singular $h \in F^*$ meets a given compact subset G_0 of E, whenever h meets G_0.

As just remarked h is bounded from any given singular point of F, so that it is permissible to suppose that G_0 is a compact subset of G. With this understood there is a finite set of right N, say $N_i (i=1,\ldots,n)$ in whose union G_0 lies. The remainder of the proof is similar to that of Lemma 7.2.

Lemma 7.1 can be extended as follows.

LEMMA 7.4. If $G^* = E$ no open subarc β of an $h \in F^*$ can have its end points on a transversal of a right N.

Suppose β existed contrary to Lemma 7.4. We can suppose that $\bar{\beta} \cap \lambda$ reduces to the end points p, q of β in λ. Let μ be a subarc of λ with the end points p and q, taking μ as the empty set when $p = q$. Then $\bar{\beta} \cup \mu$ will bound a Jordan region $R \subset E$. The number of $\alpha \in F$ incident with singular points of F is at most countably infinite so that there must be a point in R meeting a non-singular $k \in F^*$. Since R is compact it follows from Lemma 7.3 that k must meet $\bar{\beta}$ or μ. But k cannot meet $\bar{\beta}$ since k is non-singular. There must then exist an F-vector on k with end points p_1 and q_1 on μ. This is contrary to Lemma 7.1 and Lemma 7.4 follows.

COROLLARY 7.1. When $G^* = E$;there are no top circles in F^*.

COROLLARY 7.2. When $G^* = E$ there is a first and last point in which any sensed $h \in F^*$ meets the closure of an F-neighborhood, X, whenever h meets X.

COROLLARY 7.3. When $G^* = E$ there is a first a and last point in which any sensed $h \in F^*$ meets a compact subset G_0 of G^*, whenever h meets G_0.

The proof is similar to that of Lemma 7.2 on covering G_0 with a finite set of F-neighborhoods.

COROLLARY 7.4. When $G^* = E$ an h and k in F^* with $h \neq k$ intersect, if at all, in an arc, or a point, or an open arc closed at one end point.

Suppose that $h \cap k \neq 0$. It follows from Corollary 7.1 that $h \cap k$ is connected both on h and on k. Since h and k are closed in E, $h \cap k$ must be closed in E and Corollary 7.4 follows. One can add that $h \cap k$, if not empty or a point, is the union of a finite or infinite set of $\alpha \in F$ with any two successive $\alpha \in F$ joined at a common limiting end point.

From Lemma 7.2 and Corollary 7.1 one infers the following.

THEOREM 7.1. When $G^* = E$ each $\alpha \in F$ is the homeomorphic image in G of an open interval and has limiting end points in the complex z-sphere S. These end points are distinct unless coincident with $z = \infty$. Each finite end point is a singular point of F.

From Corollary 7.1 and Corollary 7.3 one infers the following.

THEOREM 7.2. When $G^* = E$ each $h \in F^*$ has limiting end points at $z = \infty$ in the complex z-sphere.

Right and left continuations of an $\alpha \in F$. A sensed $\beta \in F$ is is termed a right (left) continuation of a sensed $\alpha \in F$ if the initial point p of β $(p \in G^*)$ is the terminal point of α, and if in any F-neighborhood of p the terminal open subarc of α and the initial open subarc of β appear in a clockwise (counter-clockwise) order on the boundary of an F-sector σ incident with p. We say that σ belongs to $\alpha\beta$. If a sensed $\beta \in F$ is a right (left) continuation of a sensed $\alpha \in F$, then α, reversed in sense, is a left (right) continuation of β, reversed in sense.

A sensed $h \in F^*$, not in F, is termed a right (left) continuation of each $\alpha \in F \mid h$ if the second of any two consecutive $\beta \in F \mid h$ is the right (left) continuation of the first. An $h \in F^*$ which on sensing becomes a right or left continuation will be said to be concave towards any of its F-sectors, i.e., the F-sectors belong to successive $\beta \in F|h$.

The following lemma is a consequence of Theorem 7.1 and of the fact (Cor. 7.1) that when $G^* = E$ there are no top circles in F^*.

LEMMA 7.5. Suppose $G^* = E$. Then any $\alpha \in F$ which is not closed in E is an open subarc of just two concave $h \in F^*$; these two concave $h \in F^*$ intersect in $\bar{\alpha}$. The two F-rays on the boundary of an F-sector σ incident with a singular point p are on just one concave $h \in F^*$; this h is concave toward σ. Two concave $h \in F^*$ intersect, if at all, in a singular point of F or in an $\bar{\alpha} | (\alpha \in F)$.

THEOREM 7.3. When $G^* = E$, an $h \in F^*$ intersects an F-neighborhood X_p or a right N, if at all, in a single open arc.

We need concern ourselves here only with X_p when p is singular. For Lemma 7.4 implies that h intersects X_p when p is non-singular, or a right N, if at all, in a single open arc.

Suppose then that the theorem is false in that $h \cap X_p$ includes at least two disjoint maximal open arcs α' and α''. Not both α' and α'' can meet p, since $\alpha' \cup p \cup \alpha''$ would then be an open arc in $h \cap X_p$

containing α' and α'' as proper open subarcs contrary to the assumption that α' and α'' are maximal. Suppose then that α' does not meet p. Then α' is in an F-sector σ of X_p. Let $k \in F^*$ be constructed so as to contain the F-rays on the boundary of σ and be concave. Since k has coincident limiting end points on the complex z-sphere S at the point $z = \infty$, k bounds a region $R \subset E$ which includes σ since by Lemma 7.4 k cannot reenter σ. The other F-sectors belonging to k are also in R. Then h must be in R with α', since the concavity of k prohibits an intersection of h with k. Hence α'' cannot exist, since α' and α'' (if α'' existed) would intersect the transversal ray in σ. We infer the truth of the lemma.

COROLLARY 7.5. When $G^* = E$ an arc g in G^* which is the closure of a finite sequence of $\alpha \in F$ intersects the closure of an F-neighborhood X_p or of a right N, if at all, in a single arc or point.

The arc g is a subarc of an open $h \in F^*$. One can apply Theorem 7.3 to an F-neighborhood Y_p such that $Y_p \supset \bar{X}_p$, or to a right N_p such that $N_p \supset \bar{N}$. The corollary then follows from the theorem.

<u>Separating triples</u>. Suppose $G^* = E$. Let h, k, m be open non-intersecting arcs in F^*. It follows from Theorem 7.2 that E - h is the union of two disjoint simply-connected regions. Similar facts hold for E - m and E - k. The components of $G^* - h - k$ include a unique simply-connected region $|h,k|$ whose boundary is $h \cup k$. Moreover $|h, k| = |k, h|$. If no one of the three open arcs h, k, m separates the other two then $G^* - h - k - m$ has a unique component $|h, k, m|$ whose boundary is $h \cup k \cup m$, and (h, k, m) is termed a non-separating triple. It is clear that the property that (h, k, m) be separating or non-separating is independent of the circular order of h, k, m and that when (h, k, m) is non-separating $|h, k, m|$ is independent of the circular order of h, k, m.

<u>Transversal cuts of an</u> X_p. If X_p is an F-neighborhood of p, the union of p and of any two transversal rays of X_p is called a transversal cut of X_p.

LEMMA 7.6. If $G^* = E$ any three non-intersecting open arcs of F^* intersecting a transversal cut ζ of X_p (or $\bar{\zeta}$ of $\bar{X}_p$) form a separating triple.

By virtue of Theorem 7.3 each of the three given open arcs of F^*, h, k, m, meets ζ, if at all, in one and only one point. Suppose that h, k, m meet ζ in points p, q, r respectively, with q between p and r on ζ. No generality is lost in making this assumption since the property of (h, k, m) being separating is independent of the circular order of h, k, m. Now $E - k = A' \cup A''$, where A' and A'' are disjoint subsets of E, open and closed in $E - k$. If p is in A', r must be in A''; otherwise k would not separate E. Hence h is in A' and m in A'' so that k separates h from m.

The case in which h, k, m meet $\bar{\xi}$ is reducible to the above since $\overline{X}_p$ is contained in some larger F-neighborhood of p.

The proof of Lemma 7.7 is similar to that of Lemma 7.6.

LEMMA 7.7. If $G^* = E$ any three non-intersecting open arcs of F^* intersecting a transversal of a right N form a separating triple.

§8. BANDS R(N), $G^* = S - Z$

We suppose here that $G^* = S - Z$, where S is the complex z-sphere and Z the point $z = \infty$ on S. When $G^* = S - Z$ we shall understand that the <u>closure</u> or boundary of a set $A \subset G^*$ shall be taken <u>relative</u> to S.

Corresponding to an arbitrary right N let R(N) be the union of the sets $\alpha \in F$ which intersect N. We term R(N) a band.

LEMMA 8.1. If $G^* = S - Z$ each band R(N) is the top image of a set K in the (x, y)-plane of the form

$$(8.1) \qquad (-1 < x < 1) \quad (-1 < y < 1)$$

in which the open arcs, $x = \text{const}$, $-1 < y < 1$ in K, correspond to the $\alpha \in F \mid R(N)$.

Let an arbitrary point on the principal transversal λ of N be represented by its canonical coordinate u on the interval I: $-1 < u < 1$. Let F^s be a coherently sensed image of F, and let α_u be the open sensed arc in F^s meeting (u, 0) in λ. Let signed μ-length on α_u be measured from (u, 0) on λ so that the parameter $\mu > 0$, assigned a point $z \in \alpha_u$ following (u, 0), gives the μ-length of α_u from (u, 0) to z, while the parameter $\mu < 0$ assigned a point $z \in \alpha_u$ preceding (u, 0) gives the negative of the μ-length of α_u from z to (u, 0). Let $\mu'(u) < \mu < \mu''(u)$ be the range of μ on α_u. Making use of Lemma 3.1 it is easy to show, as in the proof of Lemma 4.1, that the function $-\mu'$ and μ'' are lower semi-continuous over I with $\mu'(u) < 0$ and $\mu''(u) > 0$. Consider first the points on α_u at which $\mu \geq 0$.

Since μ'' is lower semi-continuous there exists a sequence of positive continuous functions φ_n, defined over I, such that $\varphi_n(u) < \varphi_{n+1}(u)$,

$n = 1,2,\ldots$, while $\varphi_n(u) \to \mu''(u)$ as $n \uparrow \infty$ [3, p. 162]. Let a point $z \in \alpha_u$ with parameter μ be assigned the coordinates (u, μ). For convenience define φ_0 by setting $0 = \varphi_0(u) \mid (u \in I)$. For each $u \in I$ the set of points $z \in \alpha_u$ with coordinates (u, μ) such that

$$\varphi_{n-1}(u) \leqq \mu \leqq \varphi_n(u) \quad (n = 1,2,\ldots) \tag{8.2}$$

is mapped onto the line segment in K on which

$$x = u;\ 1 - \frac{1}{n} \leqq y \leqq 1 - \frac{1}{n+1}, \tag{8.3}$$

with y an increasing linear function of μ. There results a homeomorphic mapping of $R(N) \mid (\mu \geqq 0)$ onto $K \mid (y \geqq 0)$. In particular the point $(u, 0)$ on λ corresponds to the point $(x, y) = (u, 0)$ on the x-axis in K. One can similarly map $R(N) \mid (\mu \leqq 0)$ onto $K \mid (y \leqq 0)$ preserving the preceding mapping of points of λ. The resultant mapping of $R(N)$ onto K satisfies the lemma.

It is necessary to describe the boundary $\beta R(N)$ of $R(N)$.

LEMMA 8.2. The condition $G^* = S - Z$ implies the following.

(1) If $\beta R(N)$ contains a non-singular point p it contains the open arc $\alpha_p \in F$ meeting p.

(2) If p is a singular point of F at most one of the F-rays incident with p is in $R(N)$.

(3) If a non-singular point p is in $\beta R(N)$ and if N_p is a sufficiently restricted right neighborhood of p then one of the two components of $N_p - \alpha_p$ is in $R(N)$ the other in $S - R(N)$.

(4) If a singular point p is in $\beta R(N)$ and if X_p is sufficiently restricted, $R(N)$ contains each F-sector of X_p which meets $R(N)$.

(5) If a singular point p is in $\beta R(N)$ and if X_p is sufficiently restricted then $X_p \cap R(N)$ is either one F-sector of X_p or the union of two F-sectors of X_p and their common boundary F-ray.

(1) No point of the open arc α_p is in $R(N)$, by virtue of the definition of $R(N)$. There is however a sequence of points $p_n \in R(N)$ tending to p as $n \uparrow \infty$. Each $\alpha_n \in F$ meeting p_n is in $R(N)$. It follows from Lemma 3.1 that each point of an F-vector on α_p containing p is a limit point of Union α_n. Hence α_p itself is in $\beta R(N)$.

(2) If (2) were false there would exist an α and a disjoint $\beta \in F$ incident with p and in $R(N)$. Then $g = \alpha \cup p \cup \beta$ would have

disjoint intersections $\alpha \cap N$ and $\beta \cap N$ with N, contrary to Corollary 7.5.

(3) Let N' be a right neighborhood of p. We shall show first that there are at most two maximal open arcs in $F|(N' \cap \beta R(N))$.

Let a be one such open arc. There is a unique $\alpha \in F$ containing a. This α is in $\beta R(N)$ by (1) and is in an $h \in F^*$. This h can be constructed so that $h \cap R(N) = 0$ by virtue of (2). If then a, b, c were three maximal open arcs in $F|(N' \cap \beta R(N))$ there would exist h, k, m in F^* containing a, b, c, respectively, not meeting $R(N)$, nor each other, by virtue of their intersection with N' and Theorem 7.3. Then (h, k, m) would form a separating triple (Lemma 7.7). This is impossible since $\beta R(N)$ meets h, k, m and $R(N)$ is connected.

Hence at most two of the open arcs a, b, c can exist, and one can readily restrict a right N_p in N' so that (3) holds.

(4) Let H be an F-sector incident with p such that p is a limit point of points in $H \cap R(N)$. Let λ be a transversal of H with p as limiting end point. Then p is a limit point of points of $\lambda \cap R(N)$. Making use of the separating properties of elements in F^*, as in the proof of (3), one sees that there is at most one point of $\beta R(N)$ in λ. Hence an F-neighborhood X_p can be so restricted that $(X_p \cap \lambda) \subset R(N)$, and (4) follows.

(5) It follows from (4) and (2) that a sufficiently restricted X_p is such that $X_p \cap R(N)$ is the union of F-sectors of X_p and at most one F-ray incident with p. If then $R(N)$ included more than two F-sectors incident with p, $R(N)$ would include at least one F-sector $H \subset X_p$ with bounding F-rays a and b not in $R(N)$. It follows from (2) that $Cl(a \cup b)$ (Cl = closure) can be extended to form an $h \in F^*$ not in $R(N)$. This h would separate H from the other sectors of $X_p \subset R(N)$, contrary to the connectedness of $R(N)$. Hence there can be at most two F-sectors of $X_p \cap R(N)$ incident with p. Two such F-sectors would have an F-ray c as common boundary, and c would be in $R(N)$; otherwise a continuation of c in F^*, not in $R(N)$, would separate the two F-sectors.

Concavity or semi-concavity towards W. Let $h \in F^*$ be on the boundary of a connected region $W \subset G^*$. Then h is termed concave towards W if W contains just one F-sector incident with each singular point of F in h. We term h semi-concave towards W if W includes just one F-sector incident with each singular point of F in h except for one singular point p of F in h; corresponding to p, W shall contain just two F-sectors, incident with p and the F-ray incident with p between the two F-sectors. A singular point $p \in \beta W$ is termed of type 1 or type 2 relative to W if W contains one or two F-sectors, respectively, incident with p.

THEOREM 8.1. (a) If $G^* = S - Z$, $\beta R(N)$ is the union of Z and non-intersecting open arcs $h \in F^*$ concave or semi-concave towards $R(N)$.

(b) If such an h intersects βN, it is concave towards $R(N)$.

(c) The number of $h \in F^*$ in $\beta R(N)$ with diameters exceeding a prescribed constant $d > 0$ is finite, and the total number is countable.

It follows from Lemma 8.1, (1), (3), (5) that $\beta R(N)$ is the union of Z and of the closures of a set of elements in F, each in an $h \in F^*$ in $\beta R(N)$ whose singular points are all of type 1 or 2 relative to $R(N)$. We shall prove the following.

(i) An $h \in F^*$ in $\beta R(N)$ cannot include two F-singular points p and q of type 2 relative to $R(N)$.

If (i) were false let ζ be the subarc of h bounded by p and q. Let α and β be the open arcs in $F|R(N)$ bearing F-rays incident with p and q respectively. The open arc $\alpha \zeta \beta$ intersects N in $\alpha \cap N$ and $\beta \cap N$ contrary to Corollary 7.5. Hence (i) holds and (a) follows.

Proof of (b). If an $h \in F^*$ in $\beta R(N)$ intersects βN it contains no F-singular point q of type 2 relative to $R(N)$. To see this let $\alpha \in F$ be in h and intersect βN, and let γ in $F|R(N)$ have an end point at q (if possible). Let ζ be the arc of h between α and q (if any exists). Then $\alpha \zeta \gamma$ intersects $\overline{N}$ at least in $\alpha \cap \beta N$ and $\gamma \cap N$ contrary to Corollary 7.5. If then h intersects βN, it is concave towards $R(N)$.

Proof of (c). If the final statement of the theorem were false there would exist a sequence of points, $p_1, p_2, \ldots$ in $S - Z$, on different open arcs of F^*, $k_1, k_2, \ldots$ respectively, and such that p_n converges to a non-singular point q. Let N_q be a right neighborhood of q. For n exceeding a sufficiently large integer m, p_n will be in N_q, and hence k_n will intersect the principal transversal of N_q. In accordance with Lemma 7.7, $(k_{m+1}, k_{m+2}, k_{m+3})$ will be a separating triple. This however is impossible since k_{m+i}, $i = 1,2,3$, is in $\beta R(N)$ and $R(N)$ is connected.

We infer the truth of the concluding statement of the theorem.

§9. F-REGIONS $G^* = S - Z$

We shall need a covering of G^* by a countable number of specially selected bands $R(N)$ [cf. 7, I].

Let p be a singular point of F of exponent n and X_p an F-neighborhood of p. Let points $p_r (r = 1,\ldots,2n)$ be selected on the respective F-rays in X_p incident with p, and let M_r be a right neighborhood of p_r. The union of the bands $R(M_r)(r = 1,2,\ldots,2n)$ with p clearly yields a neighborhood of p. Hence there exists a countable sequence $Q_1, Q_2, \ldots$ of right neighborhoods such that

$$\text{Union } R(Q_r) = G \quad (r = 1,2,\ldots)$$

and such that, if G_0 is any compact subset of G^*, $G_0 - \omega$ is included in the union of a finite subset of the bands $R(Q_r)$.

LEMMA 9.1. If $G^* = S - Z$ there exists a sequence of right neighborhoods V_r, $R = 1,2,\ldots < r_0$, where r_0 is a positive integer or aleph null, such that

$$\text{(9.0)} \qquad \text{Union } R(V_r) = G \quad (r = 1,2,\ldots < r_0)$$

while $R(V_{n+1})$ $(n+1 < r_0)$ intersects the union of its predecessors $R(V_1),\ldots,R(V_n)$, but is not included in this union, and if G_0 is any compact subset of G^*, $G_0 - \omega$ is included in the union of a finite subset of the $R(V_r)$.

Note. For the present we admit the possibility that the range of r in (9.0) may be finite because of the condition that $R(V_{n+1})$ be not included in the union of its predecessors. The fact that the F-regions Σ_r whose union is G^* never have null boundaries in G^* for finite r shows that the range of r cannot be finite. See Theorem 9.1.

Set $V_1 = Q_1$. Proceeding inductively suppose $V_1, V_2, \ldots, V_n$ chosen. Then take V_{n+1} as the first of the Q_i (if any exist) such that $R(Q_i)$ intersects Union $R(V_r)$, $r = 1,\ldots,n$, but is not contained in that union. The resulting sequence $V_1, V_2, \ldots,$ finite or infinite, satisfies the lemma.

We shall refer to the interior of the closure of a point set A as Int-Cl A.

Bands $R(N)$ are special cases of F-regions Σ in G^*, characterized as follows.

(i) When Σ contains $p \in G$ it contains the $\alpha_p \in F$ which meets p.

(ii) When $\beta\Sigma$ contains $p \in G^*$ then for some X_p, $X_p \cap \Sigma$ is the Int-Cl of a proper subset of the F-sectors in X_p incident with p and consecutive about p.

(iii) Σ is open and simply-connected.

Bands $R(N)$ are F-regions. They satisfy (i) by definition, (ii) by (3) and (5) of Lemma 8.2, and (iii) by Lemma 8.1. When $G^* = S - Z$, the boundary of an F-region is clearly the union of Z and a set of non-intersecting open arcs in F^* at most countable in number, with the property that the subset of these boundary elements exceeding a prescribed constant d in diameter is finite in number. Cf. the proof of Theorem 8.1.

LEMMA 9.2. Let $G^* = S - Z$ and let Σ be an F-region in G^*. If a principal transversal λ of a right N intersects Σ, then $\bar{\lambda} \cap \bar{\Sigma}$ is a closed subarc of $\bar{\lambda}$.

$\bar{\Sigma} \cap \bar{\lambda}$ contains at least one maximal closed subarc of λ, say k, since λ is an open arc, Σ is open, and $\lambda \cap \Sigma \neq 0$. Understanding that λ is sensed, let p and q be the initial and final points of k on $\bar{\lambda}$. If $k = \bar{\lambda}$ the lemma is satisfied. If $k \neq \bar{\lambda}$ at least one of the points p and q, say q, is an inner point of λ.

The open arc α of F meeting q is in $\beta\Sigma$ and can be continued (by virtue of property (ii) of F-regions) so as to be included in an open arc h in F^* not meeting Σ. Then h separates S - Z and Σ lies on that side of h which contains p. Moreover h meets $\bar{\lambda}$ only at q (Cor. 7.5) so that there is no point $r \in \bar{\lambda} \cap \beta\Sigma$ following q on λ.

There can similarly be no point in $\bar{\lambda} \cap \beta\Sigma$ preceding p, and the lemma follows.

LEMMA 9.3. If $G^* = S - Z$ then

$$\text{Int-Cl}\,[\underset{i \leq n}{\text{Union}}\, R(V_i)] = \Sigma_n \qquad (n = 1,2,\ldots < r_0)$$

is an F-region.

The proof will be inductive. It is clear that $\Sigma_1 = R(V_1)$ is an F-region. We shall assume that Σ_{n-1}, $n > 1$, is an F-region and prove that Σ_n is an F-region.

By choice of $V_1, V_2, \ldots, V_n$, $R(V_n) | (n < r_0)$ intersects Σ_{n-1} but is not included in Σ_{n-1}. Hence the principal transversal λ_n of V_n intersects Σ_{n-1}; for Σ_{n-1} includes some point $p \in R(V_n)$ and hence includes the $\alpha_p \in F$ meeting p; every such α_p in $R(V_n)$ intersects λ_n by definition of $R(V_n)$. Hence $\lambda_n \cap \Sigma_{n-1} \neq 0$, and it follows from Lemma 9.2 that $\bar{\lambda}_n \cap \bar{\Sigma}_{n-1}$ is a closed subarc k_n of $\bar{\lambda}_n$. Then $\lambda_n - k_n$ must either be an open arc h_n (Case I), or the union of two open arcs h_n' and h_n'' (Case II), such that h_n', k_n, h_n'' appear on λ_n in the order written. We shall determine the nature of $\Sigma_n - \Sigma_{n-1}$.

Case I. Let N_n be a right neighborhood in which h_n is the principal transversal. Introduce the point $r_n = k_n \cap \bar{h}_n$. Then r_n is an inner point of λ_n. Hence the open arc $\alpha_n \in F$ meeting r_n is in $R(V_n)$. The point r_n, and hence α_n are in $\beta\Sigma_{n-1} \cap \beta R(N_n)$. Let a_n and b_n be elements in F^* which contain α_n and are in $\beta\Sigma_{n-1}$ and $\beta R(N_n)$ respectively. Then $a_n \cap b_n$ is connected, and

$$\beta\Sigma_{n-1} \cap \beta R(N_n) = (a_n \cap b_n) \cup Z. \tag{9.1}$$

For $G^* - a_n$ is the union of two components in one of which Σ_{n-1} lies, in the other $R(N_n)$, and similarly with $G^* - b_n$.

Let η_n be the maximal open subarc of $a_n \cap b_n$. Then

$$R(N_n) \cap \Sigma_{n-1} = 0 \tag{9.2}$$

$$\bar{R}(N_n) \cap \bar{\Sigma}_{n-1} = \bar{\eta}_n \cup Z \tag{9.3}$$

$$\Sigma_n = \text{Int-Cl}\,[\Sigma_{n-1} \cup R(N_n)] = \Sigma_{n-1} \cup R(N_n) \cup \eta_n \tag{9.4}$$

That Σ_n has the properties (i), (ii) and (iii) of an F-region can now be verified.

(i) Each set in the right member of (9.4) has the property (i) including η_n, so that Σ_n has the property (i).

(ii) Points $p \in \beta\Sigma_n$ which are not in the intersection (9.1) satisfy (ii), since Σ_{n-1} and $R(N_n)$ are F-regions. Points $p \in \bar{\eta}_n$ satisfy (ii) by (9.4).

(iii) Both $R(N_n)$ and Σ_{n-1} are simply-connected. They are joined in (9.4) along a simple open arc η_n which is itself added to form Σ_n. Hence Σ_n is simply-connected.

Case II. This case can be avoided entirely by replacing $R(V_n)$ in Case II by two bands $R(V_n')$ and $R(V_n'')$ which come under Case I. More explicitly let p' and p'' be the initial and final point of the closed arc $k_n = \bar{\lambda}_n \cap \bar{\Sigma}_{n-1}$ of λ_n. Then replace λ_n by the open arcs,

$$\lambda_n' = (h_n' \cup k_n) - p'' \quad \lambda_n'' = (h_n'' \cup k_n) - p'$$

whose union is λ_n, and replace V_n by successive right neighborhoods V_n' and V_n'' in which λ_n' and λ_n'' are principal transversals. Then

$$\Sigma_{n-1} \cup R(V_n) = \Sigma_{n-1} \cup R(V_n') \cup R(V_n'').$$

On setting

$$\Sigma_n' = \text{Int-Cl}\,[\,\Sigma_{n-1} \cup R(V_n')]$$
$$\Sigma_n'' = \text{Int-Cl}\,[\,\Sigma_n' \cup R(V_n'')]$$

one sees that Σ_n' and Σ_n'' are F-regions and that $\Sigma_n = \Sigma_n''$. Both of the new additions come under Case I.

This establishes Lemma 9.3.

<u>An entrance to</u> R(N). The open arc η_n in $\beta R(N_n)$ will be called an <u>entrance</u> to $R(N_n)$. More generally, given a right N, an open arc η in $\beta R(N)$ will be called an entrance to R(N) if η is the union of a finite or countably infinite set of $\alpha \in F$ each joined to its successor or predecessor at a singular point in F, and if η intersects βN. One sees that an entrance to N is an open subarc of an $h \in F^*$ in $\beta R(N)$ which intersects βN and is accordingly concave towards R(N).

If one supposes, as is possible, that each V_n comes under Case I, and in the construction of G^* out of bands replaces V_n by N_n so that (9.4) holds, one obtains a major decomposition theorem.

THEOREM 9.1. When $G^* = S - Z$ there exists a sequence of non-intersecting bands $R(N_r)$, $r = 1,2,\ldots$, and for each $R(N_r)$ with $r > 1$ an entrance η_r, such that the η_r do not intersect each other or any of the bands and such that $G^* = \text{Union } \Sigma_n$, $n = 1,2,\ldots$, where $\Sigma_1 = R(N_1)$ and for $n = 2,3,\ldots$

$$\Sigma_n = \Sigma_{n-1} \cup R(N_n) \cup \eta_n, \bar{\Sigma}_{n-1} \cap \bar{R}(N_n) = \bar{\eta}_n \cup Z \tag{9.5}$$

and where the N_r are so chosen that any compact subset of G^* is included in Σ_n for n sufficiently large.

We observe that Σ_n is an <u>F-region</u> and that

$$\beta\Sigma_n = [\beta\Sigma_{n-1} \cup \beta R(N_n)] - \eta_n \neq 0.$$

Hence Σ_n cannot coincide with G^* for any finite n. This means that in the range $0 < r < r_0$ of r in Lemmas 9.1 and 9.3 r_0 is aleph null.

§10. THE EXISTENCE OF PH FUNCTIONS

Certain preliminary lemmas are needed.

Lemma 10.1. If U is PH at a point $q \in E$ and if φ is a top mapping leaving q fixed of a neighborhood of q onto a neighborhood of q then $U\varphi$ is PH at q.

This follows almost immediately from the definition of a function PH at q.

LEMMA 10.2. If U is continuous in the rectangle $K : (-1 < x < 1)(-1 < y < 1)$ and has the level lines of x; if moreover U is PH and positive in $K|(x > 0)$ and PH and negative in $K|(x < 0)$ then U is PH over K.

Set $z = x + iy$. The values $U(z)|(z \in K)$ depend only on x. On setting $f(x) = U(z)$ we see that $f(x)$ varies continuously and strictly monotonically with x. Set $f(x) + iy = \varphi(z)$. Then φ is a homeomorphic mapping of K into a w-plane. If $U_1(w) = \mathcal{R}w$, $U = U_1\varphi$ so that U is PH over K.

LEMMA 10.3. If U is continuous in a spherical neighborhood W of the origin and PH except at most at $z = 0$, and if U has the level sets of $\mathcal{R}z^n$, $n > 0$, then U in PH in W.

Set $V(z) = \mathcal{I}z^n$ and $\varphi(z) = U(z) + iV(z)$. Then one sees that φ is 1-1 and continuous in some neighborhood of each point $z_0 \neq 0$ in W. For $z \neq 0$, φ is either sense preserving or else its conjugate $\bar{\varphi}$ is sense preserving. Set $f = \varphi$ or $\bar{\varphi}$ according as φ or $\bar{\varphi}$ is sense preserving for $z \neq 0$. Then f is interior over $W|(z \neq 0)$ and in accordance with Lemma 2.1 [5] interior over W. Set $U_1(w) = \mathcal{R}w$. Then $U(z) = U_1f(z)$ so that U is PH over W.

THEOREM 10.1. If $G^* = S - Z$ there exists a function U, PH over G^* with the open arcs of F as level arcs and the points of ω as critical points.

Let F^s be a coherently sensed image of F. We shall refer to Theorem 9.1 and define U on $\Sigma_1, \Sigma_2, \ldots$ in such a manner that $U|\Sigma_{n+1}$ is an extension of $U|\Sigma_n$.

<u>Definition of</u> $U|\Sigma_1$. Let the point on the principal transversal h_1 of N_1 with canonical coordinates $(u, v) = (u, 0)$ be represented by $h_1(u)$. Let $\beta_u \in F^s$ meet $h_1(u)$. Set

$$U(z)|(z \in \beta_u) = u. \tag{10.2}$$

We shall show that U so defined is PH over Σ_1.

Let signed μ-length on β_u be measured from $h_1(u)$. The point z on β_u with parameter μ has a representation $z = X(u, \mu)$ which defines a top mapping of the domain

$$(\mu'(u) < \mu < \mu''(u)) \quad (-1 < u < 1) \tag{10.3}$$

onto Σ_1, where $(\mu'(u), \mu''(u))$ gives the range of μ on β_u. Set $w = u + i\mu$ and write the inverse of $z = X(u,\mu)$ in the form $w = \varphi(z)$. If $U_1(w) = \mathcal{R}w$ then $U = U_1\varphi$ over Σ_1 so that U is PH over Σ_1. Note also that $U(z)$ is bounded for $z \in \Sigma_1$.

<u>The induction</u>. We assume that $U|\Sigma_{n-1}$ has been defined for $n > 1$, is bounded and PH over Σ_{n-1}, and is constant on the open arcs of $F|\Sigma_{n-1}$, and shall show that U can be extended in definition over Σ_n with like properties.

<u>Definition of</u> $U|R(N_n)$. We refer to the transversal h_n of N_n of the proof of Lemma 9.3. Let h_n be extended slightly beyond its end point $r_n \in \beta\Sigma_{n-1}$ so as to enter Σ_{n-1} as a transversal λ_n. Let $k_n = \lambda_n - h_n$ in $\bar{\Sigma}_{n-1}$ be parameterized in the form $z = k_n(t)$ by assigning to $z \in k_n$ a parameter value t equal to $U(z)$ as defined in Σ_{n-1}. This representation of k_n is 1-1 and continuous. Let λ_n, including k_n, now be parameterized as an extension of $k_n(t)$, making this parameterization $\Lambda_n(t)$ 1-1 and continuous and enlarging the range of the parameter t by an arbitrary positive constant c_n. For $\Lambda_n(t) \in R(N_n)$ and a $\gamma_t \in F^s$ meeting $\Lambda_n(t)$ set $U(z)|(z \in \gamma_t) = t$. This completes the definition of U over $R(N_n)$.

<u>Definition of</u> $U|\eta_n$. To define $U|\eta_n$ we first show that $U(z)$ has a unique limit at r_n as r_n is approached as a limit either from Σ_{n-1} or from $R(N_n)$. If N is a sufficiently restricted right neighborhood of r_n and if $\alpha_n \in F^s$ meets r_n, one component K_1 of $N - \alpha_n$ is in Σ_{n-1} and the other component does not intersect Σ_{n-1}. If u and v are canonical coordinates in K_1 of z then $U(z)|(z \in K_1)$ has a value $\psi(u)$ which is strictly monotone and bounded. Hence $U(z)$ tends to a limit c as $z \to r_n$ in Σ_{n-1}. Similarly $U(z)$ tends to a limit c' as $z \to r_n$ from $R(N_n)$. But r_n lies on the extended transversal λ_n, and divides λ_n into a transversal in Σ_{n-1} and a transversal in $R(N_n)$. At a point $z = \Lambda_n(t) \neq r_n$ of λ_n, whether in Σ_{n-1} or in $R(N_n)$, $U(z)$ is defined and equals t. Hence $c = c' = \Lambda_n^{-1}(r_n)$.

We accordingly complete the definition of $U|\Sigma_n$ by setting $U(z)|(z \in \eta_n) = c = c'$. Observe that $U(z) - c$ has opposite signs in $\Sigma_{n-1} \cap N$ and $R(N_n) \cap N$. Note also that U is bounded over Σ_n.

<u>Proof that</u> $U|\Sigma_n$ <u>is</u> PH. As defined, U is PH over $\Sigma_{n-1}|(n > 1)$ by hypothesis, and over $R(N_n)$, as we have shown for $R(N_1) = \Sigma_1$. It remains to show that U is PH at each point $p \in \eta_n$.

If p is non-singular we introduce a right neighborhood M_p with canonical coordinates (u, v). Suppose first $p \in \alpha_n$, where $\alpha_n \subset \eta_n$ is defined above. Without loss of generality we can suppose that $U(p) = 0$ and that U is positive in M_p when $u > 0$ and negative when $u < 0$. Moreover $U|M_p$ is continuous. It follows from Lemmas 10.2 and 10.1 that U is PH over $\Sigma_{n-1} \cup R(N_n) \cup \alpha_n$.

If $\alpha_n \neq \eta_n$ let p be an end point of α_n and let α'_n join α_n in η_n at p. Then $\Sigma_{n-1} \cup R(N_n) \cup \alpha_n \cup \alpha'_n \cup p$ includes a neighborhood X_p of p. It follows from the definition of U on the components of this union that $U|X_p$ is continuous and that U changes sign on crossing α'_n. It follows from Lemmas 10.2 and 10.1 that U is PH at each point of α'_n. Continuing thus with successive $\alpha \in F|\eta_n$ we see that U is continuous over Σ_n and PH at each point of each $\alpha \in F|\eta_n$.

If $p \in \eta_n$ is <u>singular</u> we refer to an F-neighborhood X_p with complex parameter $w = u + iv$ in terms of which $U(z)$ has the values $Y(w)$ and the level sets of $\mathcal{R} w^n$. By virtue of the result just established for non-singular points of η_n, Y is PH in a neighborhood of $w = 0$ except at most at $w = 0$. It follows from Lemma 10.3 that Y is PH at $w = 0$. It follows then from Lemma 10.1 that U is PH at $z = 0$.

Thus U is PH on Σ_n without exception. Since Union $\Sigma_n = G^*$ the proof of the theorem is complete.

BIBLIOGRAPHY

[1] BOOTHBY, W. M., The topology of regular curve families with multiple saddle points. American Journal of Mathematics, vol. 73 (1951), pp. 405-438.

[2] BOOTHBY, W. M., The topology of level curves of harmonic functions with critical points. American Journal of Mathematics, vol. 73 (1951), pp. 512-538.

[3] CARATHÉODORY, C., Reelle Funktionen. I. Leipzig and Berlin, 1939.

[4] JENKINS, J. A., On the topological theory of functions. Canadian Journal of Mathematics, vol. 3 (1951), pp. 276-289.

[5] JENKINS, J. A., MORSE, MARSTON, Contour equivalent pseudoharmonic functions and pseudoconjugates. American Journal of Mathematics, vol. 74 (1952), pp. 23-51.

[6] JENKINS, J. A., MORSE, MARSTON, The existence of a pseudoconjugate on Riemann surfaces. To appear.

[7] KAPLAN, WILFRED, Regular curve families filling the plane. Duke Mathematical Journal. I. vol. 7 (1940), pp. 154-185; II. vol. 8 (1941), pp. 11-45.

[8] KAPLAN, WILFRED, Differentiability of regular curve families on the sphere. Lectures in Topology, Ann Arbor, 1941, pp. 299-301.

[9] KAPLAN, W., Topology of level curves of harmonic functions. Transactions of the American Mathematical Society, vol. 63 (1948), pp. 514-522.

[10] MORSE, MARSTON, Topological methods in the theory of functions of a complex variable. Annals of Mathematics Studies, Princeton, 1947.

[11] MORSE, M., HEINS, M., Topological methods in the theory of functions of a single complex variable. Annals of Mathematics. I. (1945), pp. 600-624; II. (1945), pp. 625-666; III. (1946), pp. 233-274.

[12] MORSE, M., HEINS, M., Deformation classes of meromorphic functions and their extensions to interior transformations. Acta Mathematica, vol. 79 (1947), pp. 51-103.

[13] STOÏLOW, S., Lecons sur les principes topologiques de la théorie des fonctions analytiques. Paris, 1938.

[14] TÔKI, YUKANARI, A topological characterization of pseudoharmonic functions. Osaka Mathematical Journal, vol. 3 (1951), pp. 101-122.

[15] WHYBURN, G. T., Analytic topology. American Mathematical Society Colloquium Lectures, New York, 1942.

COVERINGS OF RIEMANN SURFACES

Léonce Fourès

A. RIEMANNIAN COVERINGS

§1. COVERINGS

A surface is a triangulable, and here orientable, two-dimensional manifold. Neighborhoods will always be homeomorphic to the open circle. Covering surfaces have been defined and used in topology and the theory of functions [1], [3], [5], [6], [8], [14] ... . Some authors use two corresponding triangulations on the given surface and on its covering [3], [6]. We will give here local definitions, similar to those given by Stoïlow [8] and C. Chevalley [14].

In this part t will be a continuous mapping of a surface $\mathcal{S}$ into another surface S, such that if $\mu \in S$, $t^{-1}(\mu)$ is an isolated set (consists of isolated points).

DEFINITION 1. $P \in S$ is covered without ramification by $\Delta \subseteq \mathcal{S}$ (with respect to t), if there exists a neighborhood $V(P) \subset S$ such that $E_{V,\Delta} = \Delta \cap t^{-1}[V(P)] = \emptyset$ and that t maps every component of $E_{V,\Delta}$ biuniquely onto $V(P)$. $(\mathcal{S}, t)$ is a relatively non ramified covering of S if every $P \in S$ is covered without ramification by $\mathcal{S}$ (with respect to t). [1]

According to the first part of this definition there exists in every component Γ_1 of $E_{V,\Delta}$ a point $\mathcal{P}_1$ such that $t(\mathcal{P}_1) = P$. From the second part we see that t is a mapping of $\mathcal{S}$ onto S.

SPECIAL CASE. Suppose S is of genus zero, doubly connected. Let M S. If $t^{-1}(M)$ has a finite number of points in $\mathcal{S}$, then $\mathcal{S}$ is doubly connected and n, which is independent of $M \in S$, is the multiplicity of the covering. If t^{-1} has

(1) We shall say sometimes that $\mathcal{S}$ is a covering of S when the choice of the function which realizes the mapping of $\mathcal{S}$ onto S cannot be misunderstood.

an infinite number of points in $\mathscr{S}$, then $\mathscr{S}$ is simply connected and so $(\mathscr{S}, t)$ is the universal covering of S.

DEFINITION 2. $P \in S$ is covered with ramification by $\Delta \subseteq \mathscr{S}$ (with respect to t) if there exists a $V(P) \subset S$ such that

(i) $E_{V,\Delta} \neq \emptyset$

(ii) If Γ_1 is any component of $E_{V,\Delta}$, then $\mathscr{P}_1 = \Gamma_1 \cap t^{-1}(P) \neq \emptyset$

(iii) $\Gamma_1^* = \Gamma_1 - \mathscr{P}_1$ is a relatively unramified covering of $V^*(P) = V(P) - P$, with finite multiplicity.

$(\mathscr{S}, t)$ is a ramified covering of S if any $P \in S$ is covered with or without ramification by (with respect to t).

From (iii) and the fact that $V^*(P)$ is doubly connected, it follows that Γ_1^* is doubly connected and $\mathscr{P}_1$ consists of a unique point, since $\mathscr{P}_1$ is already an isolated set.

DEFINITION 3. $P \in S$ is regularly covered by $\mathscr{S}$ (with respect to t) if there exists a $V(P) \subset S$ such that all the components Γ_1 of $E_{V,\mathscr{S}}$ cover P with the same multiplicity

If every $P \in S$ is regularly covered by $\mathscr{S}$ (with respect to t), $(\mathscr{S}, t)$ is called a regularly ramified covering of S.

t is single-valued on $\mathscr{S}$, but t^{-1} is not single-valued on $S^* = t(\mathscr{S})$. Let $\mathscr{P} = t^{-1}(P)$ which defines a branch of t^{-1} in a neighborhood of P. P is a branch point with multiplicity n of this branch if there exists a V(P) such that the component of $E_{V,\mathscr{S}}$ which contains $\mathscr{P}$ is a ramified covering of P, with multiplicity n.

§2. RIEMANNIAN COVERING

In definitions 1, 2, 3, let us replace the surfaces S and $\mathscr{S}$ by two Riemann surfaces R and $\mathscr{R}$, and t by a mapping q satisfying the same conditions as t and which is conformal and one-to-one wherever t was one-to-one.

Let D be a doubly connected domain of genus zero. Let (Δ, t), (δ, φ) be two relatively non ramified coverings of D, with the same multiplicity n; every branch of $\theta^{-1} \cdot t$ or $t^{-1} \cdot \theta$ is uniform. Let C be the unit circle of the z-plane; Γ_1 and Γ_2 the unit circles of the ζ_1 and ζ_2 planes. Let us denote by $C^*, \Gamma_1^*, \Gamma_2^*$ the domains obtained by removing the origins of the circles C, Γ_1, Γ_2. Let $z = q_1(\zeta_1)$ and $z = q_2(\zeta_2)$ be two functions continuous in Γ_1 and Γ_2 with $q_1(0) = q_2(0) = 0$, and such that (Γ_1^*, q_1) and (Γ_2^*, q_2) are coverings of C^* with the same multiplicity. Then $\psi = q_1^{-1} \cdot q_2$ and ψ^{-1} are single-valued with $\psi(0) = 0$ $\psi^{-1}(0) = 0$. The origin of Γ_2 is regular for ψ which then takes the form $\psi = \alpha\zeta_2$ $|\alpha| = 1$. (Γ_1^*, ζ_1^n) being a covering of C^*, with multiplicity n, all the coverings of C^* with multiplicity n, have the form $(\Gamma^*, k\zeta^n)$, whenever Γ^* is a unit circle without its origin.

Let $(\mathcal{R}, q)$ be a ramified or unramified covering of R. Let $P \in R$, $\mathcal{P} \in \mathcal{R}$ such that $q(\mathcal{P}) = P$. There exists $V(P)$ such that the component $\mathcal{V}(\mathcal{P})$ of $E_{V,\mathcal{R}}$ which contains $\mathcal{P}$ covers P with the multiplicity n. The local uniformizing functions of $\mathcal{R}$ and R, denoted by $\mathcal{T}$ and T, map $\mathcal{V}(\mathcal{P})$ and $V(P)$ conformally onto the unit circles Γ and C of the ζ-plane and the z-plane respectively. From the preceding we get $T \cdot q \cdot \mathcal{T}^{-1} = \psi = k\zeta^n$.

THEOREM 1. Let $(\mathcal{R}, q)$ be a ramified covering of R

a) The points where q is not one-to-one are isolated.

b) Let $P = q(\mathcal{P})$: one can find on R and $\mathcal{R}$ two neighborhoods $V(P)$ and $\mathcal{V}(\mathcal{P})$ such that their parameters z and ζ which are obtained by the local uniformizing functions T and $\mathcal{T}$, satisfy $T \cdot q \cdot \mathcal{T}^{-1} = k\zeta^n$.

So the definition 2 in which $R, \mathcal{R}, q$ took the place of $S, \mathcal{S}, t$ is a generalization of the Riemannian covering which was first defined by Stoilow: he mapped an abstract Riemann surface only onto the complex plane, and this was used to define abstract Riemann surfaces. We will still say that $(\mathcal{R}, q)$ is a Riemannian covering of R, q a Riemannian projection of $\mathcal{R}$ onto R.

Let us denote by $\mathcal{S}\, u$ every function of the variable u which can be written in a neighborhood of $u = 0$ as a convergent series

$$\mathcal{S}\, u = \alpha_1 u + \alpha_2 u^2 + \dots \quad \alpha_1 \neq 0 .$$

In the same way $\mathcal{S}\mathcal{S}\,(u,v)$ will denote a double series, in the variables u and v and without constant term, and which is convergent in a neighborhood of $u = 0$ $r = 0$,

$$\mathcal{S}\mathcal{S}\,(u,v) = \alpha_{11}u + \alpha_{12}v + \alpha_{21}u^2 + \alpha_{22}uv + \dots$$

If q is a Riemannian projection $M = q(\mathfrak{M})$ one gets for local uniformizing functions T and $\mathcal{T}$ which map particular neighborhoods of M and $\mathfrak{M}$: $z = T \cdot q \cdot \mathcal{T}^{-1} = k\xi^n$. If we consider the local uniformizing functions $z = T(M)$ and $\xi = \mathcal{T}(\mathfrak{M})$ for arbitrary neighborhoods of M and $\mathfrak{M}$ we can find from the previous relation

$$(1) \qquad z = \mathcal{S}(\mathcal{S}\xi)^n = \xi^n(1+\mathcal{S}\xi)$$

which is invariant under every transformation $\mathcal{S}$ of ξ and z. In (1) the functions $\mathcal{S}$ are completely arbitrary.

§3. RIEMANN SURFACES

Let $W = F(Z)$ be a mapping of a Riemann surface R_z into another R_w. This mapping is analytic if for any pair of points Z and W such that $W = F(Z)$, one can find two particular neighborhoods $V_z(Z) \subset R_z$ and $V_w(W) \subset R_w$ such that if T_z and T_w are the local uniformizing functions of R_z and R_w which map V_z and V_w onto the unit circles of the z and w plane, thus in a neighborhood of the origin

$$w = h(z) = T_w \cdot F \cdot T_z^{-1} = \alpha_0 z^{\frac{p}{q}} (1+ \alpha_1 z^{\frac{1}{q}} + \alpha_2 z^{\frac{2}{q}} + \ldots).$$

If $u = z^{\frac{1}{q}}$ $s = w^{\frac{1}{p}}$, then $s = q(u)$ where q is homomorphic, schlicht and vanishes at the origin. Thus w can be written in the form

$$w = [q(z^{\frac{1}{q}})]^p = (\mathcal{S} z^{\frac{1}{q}})^p = \alpha z^{\frac{p}{q}} (1+\mathcal{S} z^{\frac{1}{q}})$$

which is invariant under every transformation $\mathcal{S}$ of z or w. $\mathcal{S}$ is completely arbitrary.

DEFINITION 4. The Riemann surface of the analytic function $W = F(Z)$ is the set $(\mathcal{R}; \psi, \omega)$ of a Riemann surface $\mathcal{R}$ and two Riemannian projections of $\mathcal{R}$ onto R_z and R_w: $Z = \psi(M)$, $W = \ (M)$, $(M \in \mathcal{R})$ such that:

(i) $F = \omega \cdot \psi^{-1}$

(ii) If the F image in R_w of a closed curve of R_z, is closed, the ψ^{-1} image is also closed.

There exists on R_z a neighborhood $V(Z)$ such that the component Δ_v of $\psi^{-1}(V)$, which contains M, is mapped by a local uniformizing function of $\mathcal{R}$, Θ_v, onto the unit circle G_v of the ξ_v plane, such that $T_z \cdot \psi \cdot \Theta_v^{-1} = \xi_v^q$ (where T_z is a local uniformizing function of R_z which maps V onto the unit circle of the z-plane. Similarly there exists on R_w a $\mathcal{V}(w)$ such that one has with similar notations $T_w \cdot \omega \cdot \Theta_{\mathcal{V}}^{-1} = \xi^p$. Let $\Delta = \Delta_v \cap \Delta_{\mathcal{V}}$.

M belonging to Δ is mapped into $\delta_v \subset G_v$ and $\zeta_v \subset G_v$ and $\zeta_{\mathscr{V}} \subset G_{\mathscr{V}}$. Between these images there exists a conformal one-to-one correspondence $\zeta_{\mathscr{V}} = q\ (\zeta_v)$ which maps the origins into each other. Let D_z and D_w be the images of Δ by $T_z \circ \psi$ and by $T_w \circ \omega$. Notice that $\Theta^{-1}_{\mathscr{V}} \circ q \circ \Theta_v$ is the identity on $\mathscr{R}$. Then in D_z:

$$w = f(z) = T_w \circ \omega \circ \psi^{-1} \circ T_z^{-1} =$$

$$T_w \circ \omega \circ \Theta^{-1}_{\mathscr{V}} \circ q \circ \Theta_v \circ \psi^{-1} \circ T_z^{-1} =$$

$$[\varphi(z^{\frac{1}{q}})]^p \quad .$$

Let us show that if p and q have the same meaning as in the beginning, the expansion of $w = f(z)$ cannot have the form

$$w = \left(\mathcal{S}\, z^{\frac{1}{q'}}\right)^{p'}$$

with $p' < p$, $q' < q$. Otherwise there would be a closed curve in V_z which runs q' times around Z, to which there corresponds a closed curve in V_w which runs p' times around w. To these there would correspond a closed curve in $\mathscr{R}$. But ψ maps a closed curve of $\mathscr{R}$ which runs around M onto a closed curve in R_z which runs at least q times around z. Then $q' > q$ which contradicts the hypothesis.

THEOREM 2. Let $(\mathscr{R}; \psi, \omega)$ be the Riemann surface of $W = F(Z)$. The multiplicities of the Riemannian projections $Z = \psi(M)$ and $W = \omega(M)$ are respectively the coefficients q and p of the expansion $w = (\mathcal{S}\, z^{\frac{1}{q}})^p$ obtained between the parameters z and w of arbitrary neighborhoods of Z and W.

B. AUTOMORPHISM FUNCTIONS

§1. DEFINITIONS

Let $w = f(z)$ be single-valued $z' = \varphi(z)$ is an automorphism function of f if $f = f \circ \varphi$. φ can then be written $\varphi = f^{-1} \circ f$ [2], [4], [9], [11], [12], [13]. Now let $w(z)$ be an algebraic function defined by $P(w,z) = 0$, where P is a polynomial. Suppose that w has a common value for z and z'. We get a relation between z and z' by eliminating w between $P(w,z) = 0$ and $P(w,z') = 0$. If w has a common value for z and z' and a common value for z' and z", $w(z)$ does not necessarily have a common value for z and z". Indeed

$P(w_1,z) = 0$ $P(w_,\ z') = 0$	holds	$P(w_2,z') = 0$ $P(w_2,z'') = 0$	holds

where it is possible that $w_1 \neq w_2$.

In order to be able to compose automorphism functions we will define the automorphism functions in the following way.

DEFINITION. Let $(\mathcal{R}; \psi, \omega)$ be the Riemann surface of the analytic function $W = F(Z)$. $M' = \Phi(M)$ is an automorphism function of $w = \omega(M)$ if $\omega(M') = \omega(M) = \omega[\Phi(M)]$. Then $\Phi = \omega^{-1} \circ \omega$. The relation $Z' = \psi \circ \Phi \circ \psi^{-1}$ obtained between $Z = \psi(M)$ and $Z' = \psi(M')$ will be called the projection into R_z of the autormorphism function of $w = \omega(M)$.

§2. BRANCH POINTS

Let $M' = \Phi(M)$ be an automorphism function: $W = \omega(M) = \omega[\Phi(M)]$. Let us consider on the surfaces $\mathcal{R}, \mathcal{R}, \mathcal{R}_w$ neighborhoods of M, M', W and the local uniformizing functions which map them onto the unit circle of the ζ-, ζ'-, ω-planes.

$$w = a\,\zeta^{p}(1+S\zeta) = a'\,\zeta'^{p'}(1+S\zeta')$$

$$\alpha\,\zeta^{\frac{p}{p'}}(1+S\,\zeta) = \alpha'\,\zeta'(1+S\zeta) = S\zeta' .$$

If $S\zeta' = u$ then $\zeta' = S u$

$$\zeta' = S[\zeta^{\frac{p}{p'}}(1+S\zeta)] \tag{3'}$$

$$\zeta' = \gamma\,\zeta^{\frac{p}{p'}}\,[1 + SS(\zeta^{\frac{p}{p'}}, \zeta)] \tag{3}$$

In the expression (3) which is invariant under every transformation S of ζ and ζ', the double series SS is not arbitrary; (3) written as a simple series contains only terms

$$\alpha_1\,\zeta^{\frac{\lambda p + \mu p'}{p'}}$$

where $\lambda > 0$ $M \geq 0$.

Let us project $M' = \Phi(M)$ onto the surface R_z by $Z = \psi(M)$

$$w = \left(S\,z^{\frac{1}{q}}\right)^{p} = \left(S\,z^{\frac{1}{q}}\right)^{p'}$$

From this we get

$$S\,z'^{\frac{1}{q'}} = \left(S\,z^{\frac{1}{q}}\right)^{\frac{p}{p'}} \quad \text{or} \quad z'^{\frac{1}{q'}} = S\left(S\,z^{\frac{1}{q}}\right)^{\frac{p}{p'}}$$

$$z' = \left[S\left(S\,z^{\frac{1}{q}}\right)^{\frac{p}{p'}}\right]^{q'} \tag{4'}$$

$$z' = c\,z^{\frac{p}{p'}\frac{q'}{q}}\left[1 + SS\left(z^{\frac{p}{qp'}}, z^{\frac{1}{q}}\right)\right] \tag{4}$$

The expressions (4) and (4') are invariant under every transformation S of z and z'. In (4) the double series SS the coefficients are not arbitrary but (4) can be written as a simple series which contains only terms

$$a_1 z^{\frac{\lambda p + \mu p'}{qp'}}$$

where $\lambda \geq m\mu > 0$.

THEOREM 3. Let $M' = \Phi(M)$ be an automorphism function of the analytic function $W = F(Z)$. Φ is defined on the Riemann surface $\mathcal{R}$ of F. Let p be the multiplicity of $W = \omega(M)$ in M, p' the multiplicity in M', q and q' the multiplicities of $Z = \psi$ in M and M'.

Then the expansion of $M' = \Phi(M)$ has the form

$$\zeta' = \gamma \zeta^{\frac{p}{p'}} \left[1 + SS\left(\zeta^{\frac{p}{p'}}, \zeta\right)\right]$$

which contains only terms

$$\alpha \zeta^{\frac{\lambda p + \mu p'}{p'}} \quad \lambda > 0 \quad \mu \geq 0. \tag{3}$$

The projection of Φ by ψ, that is to say $\psi \circ \Phi \circ \psi^{-1}$ has for expansion

$$z' = c\, z^{\frac{p}{p'} \cdot \frac{q'}{q}} \left[1 + SS\left(z^{\frac{p}{qp'}}, z^{\frac{1}{q}}\right)\right] \tag{4}$$

which contains only terms

$$az^{\frac{\lambda p + \mu p'}{p'q}} \quad \lambda \geq m \quad \mu > 0.$$

The branch points of Φ and $\psi \circ \Phi \circ \psi^{-1}$ are not of the general type of branch point for an analytic function (see eq. (2)).

§3. AUTOMORPHISM FUNCTIONS WITHOUT BRANCH POINTS

Let $M' = \Phi(M)$ be an automorphism function without branch points such that its inverse also has no branch point.

$\mathcal{R}$ can be mapped onto a circle C (of finite or infinite radius) or the whole plane by the global uniformizing function $f \cdot f^{-1}$ is invariant under the linear functions of a properly discontinuous group G. Let ϕ_1 be any branch of Φ, (defined in a neighborhood of a point). Thus $g_1 = f \circ \phi_1 \circ f^{-1}$ is linear, so that $\omega \circ f^{-1}$ is automorphic with respect to a linear transformation $g_1 \notin G$, and to every function of the properly discontinuous group Γ obtained by adjoining to G the function g_1. The group Γ contains of course all functions $g_j = f \circ \phi_j \circ f^{-1}$. Indeed, ϕ_j

is obtained by continuation of ϕ_1 along a closed curve on $\mathcal{R}$, the image of which is open in C and connects two points which are mapped into each other by a function of G. So, $g_j = a \circ g_1 \circ b$, where a and $b \in G$, then $g_j \in \Gamma$. Conversely every ϕ_k obtained by $\phi_k = f^{-1} \circ g_k \circ f$ where $g_k \in \Gamma$ $g_k \notin G$ is a branch of Φ. The images by the same function g_j of two points of C which are mapped into each other by $a \in G$ are not always mapped into one another by a function of G, as soon as Φ has more than one branch, i.e., $g_j \circ a \circ g_j^{-1}$ is not always in G ($a \in G$).

<u>G is an invariant subgroup of Γ if and only if Φ is uniform.</u> Let n be the number of classes γ_1 defined in Γ in the following way: g_μ and g_ν belong to the same class if and only if $g_\mu \circ g_\nu^{-1} \in G$, (G is itself one of these classes). In the general case the number of branches of Φ is then equal to n-1. So Φ <u>and Φ^{-1} have the same number of branches</u>.

REMARK 1. The theorem of §3 shows that if an analytic representation $Z' = \Phi(Z)$ of a Riemann surface R_z onto itself is given, there does not always exist an analytic function $W = F(Z)$ with Φ as projection of one of its automorphism functions. But if one considers in R_z a small enough domain (in order that Φ be a one-to-one and conformal mapping of D onto $D' \subset R_z$ and that some conditions of regularity are satisfied between boundaries), one can always form a domain Δ, $\Delta \subset R_z$, $\Delta \supset D$, $\Delta \supset D'$ and a function $W = F^*(Z)$ single valued in Δ and such that in Δ, Φ is an automorphism function of F*[17].

C. REGULARLY RAMIFIED TOPOLOGICAL TREES

§1. DEFINITIONS

Let (Σ,t) be a covering of a closed surface B_0, of genus zero. Suppose this covering is relatively ramified at a finite number of points α_i (i=1,2,...,q). One can represent Σ by its topological tree T. Its sides are of q types denoted by $q_1,q_2,\dots,q_{q-1},q_q$ and are met in this order when one goes around a • knot (in the positive sense), around a x knot (in the negative sense).

We will say that T is traced on a surface S if T is on the surface S and if moreover every part of S, bounded by sides of T, and containing no side of T in its interior, is simply connected.

Given T, one can define a polygon of T as a simply connected surface bounded by a polygonal line built by sides of types q_i and a_{i+1} only. This polygonal line can be either closed, or open. In the latter case it may have an infinite number of sides.

We will consider only trees T traced on a surface S which they triangulate, that is to say trees all of whose polygons have a finite number of sides. Let us denote by K_i the polygons described on S by T. In every K_i let us choose a point Ω_i which we will call the "center" of K_i. Let $K_{i_1}\ K_{i_2}\ \ldots\ K_{i_q}$ be the polygons with N_j as common vertex. Let us connect Ω_{i_k} and $\Omega_{i_{k+1}}$ (k defined mod q): We get a polygon K^j. The K^j form a triangulation of S, by means of which one gets a topological correspondence between S and Σ. $\Sigma \hookrightarrow S$.

REMARK 2. Since (Σ, t) is a covering of B_o, it follows from our definitions that all the points on B_o are covered by Σ (with respect to t), so all the polygons of T have a finite number of sides.

We will say that *two knots A and B of T are equivalent $(A \sim B)$ if there exists a homeomorphism of T onto itself which maps A into B. T is said to be regularly ramified if there is only a finite number of classes of equivalent knots.*

A set E of knots is connected if one can join any two knots of E with a chain of sides which connects only knots of E.

In the following, we will assume that T is regularly ramified.

§2. CELLS

A cell is a finite, connected set of non equivalent knots, such that every knot of T is equivalent to some knot of the cell.

Construction. Let E be a set of knots of T : $M \notin E$ is called extra peripheral for E, if it can be connected to a knot of E by only one side of T. $N \in E$ is called peripheral for E if N can be connected to an extra peripheral knot by a unique side.

Let us choose a knot $A = A_o^{m_o} \in T$. Let $A_1^1\ A_1^2\ \ldots\ A_1^{n_1}$ be the extra peripheral knots for $A_o^{m_o}$. Let us build up C_1 as follows: first we order the set of pairs of integers i.e., $(x,y) < (x',y')$, if either $n < n'$ or $n = n'$ and $y < y'$. Then $A_1 \in C_1$ if $A_1 \not\sim A_j$ where $(j,\mu) < (1,\nu)$. Let us denote by $A_o^{m_o}, A_1^1, A_1^2, \ldots, A_1^{m_1}$ the knots of C_1 and in general by $A_o^{m_o}, A_1^1, A_1^2, \ldots, A_1^{m_2}, A_2^1, \ldots, A_2^{m_2}, A_3^1, \ldots, A_i^{m_i}$, the knots of C_i.

Let $A_{i+1}^1, A_{i+1}^2, \ldots, A_{i+1}^{n_{i+1}}$ be the extra peripheral knots for C_i.

$$A_{i+1}^{\nu} \in C_{i+1} \text{ if } A_{i+1} \not\sim A_j \text{ where } (j,\mu) < (i+1,\nu).$$

Thus every knot of C_i also belongs to C_{i+1}. But the sequence of C_i's cannot increase indefinitely since there is in T only a finite number of classes of equivalent knots. Then for some k $C_{k+1} = C_k = C$.

<u>Association of free sides of C</u>. A free side of C is a side of T which connects a peripheral knot of C to an extra peripheral knot. Let Q be extra peripheral; Q has an equivalent knot R in C. Let α be a free side of C which connects Q to a peripheral knot Q'. Then there is a side α' of the same kind as α which connects R to R'. $R' \neq R$ otherwise $R = Q$. $R' \sim Q$ implies $R' \notin C$ and R is peripheral for C. So the sides R R' (α') and Q Q' (α) are both free sides; we shall say R R' and Q Q' are associated. Obviously this association is reflexive and any free side has a unique associated side.

<u>C is a cell</u>. If a point P_1 of T is equivalent to P of C, then the homeomorphism of T onto itself which sends P into P_1 maps C onto a unique set C_1. Two sets C_1 are either identical or without any common points.

For every knot of T there is an equivalent knot in C. Let $P \in T$, one can connect P to $O \in C$ by a polygonal line L with a finite number of sides. Let us follow L (from O to P). Let P_1 be the first point on L which is extra peripheral for C. P_1 has an equivalent knot P_1' in C and one can build C_1 corresponding to C under the homeomorphism of T onto itself which sends P_1' onto P_1. Thus $C_1 \cap C = \emptyset$. Let P_2 be the first extra peripheral knot for C C_1. It is extra peripheral for at least one of the sets C, C_1 and so has an equivalent knot $P'_2 \in C$. Let us build C_2 corresponding to C under the homeomorphism of T onto itself which sends P_2' onto P_2 .... Since $P_2 \neq P_1$ and since L contains only a finite number of knots, there exists a C_n containing P. Then P has an equivalent knot in C, of course unique. The set C is a cell which we will denote by $\mathfrak{C}$.

§3. T AS A RAMIFIED COVERING.

If $\mathfrak{C}$ is not traced on a plane let us cut some of the sides in order to get a set $\mathfrak{C}^*$ traced on a plane: $\mathfrak{C}$ can be obtained from $\mathfrak{C}^*$ by identification of some of the free sides. Let us bound $\mathfrak{C}^*$ by a polygon π such that every free side of $\mathfrak{C}^*$ ends on a unique side of π. Let us pair off the sides of π according to the pairing of the free sides of $\mathfrak{C}^*$ which end on them. By identifying associated sides of π one gets a closed surface S_o and $\mathfrak{C}^*$ is mapped on a tree T_o which forms a triangulation of S_o. Let us denote by p* the mapping of T onto T_o which maps a knot of T onto the image in T_o of its equivalent in $\mathfrak{C}$. Let K be a polygon of S. Every side of K has a unique image in T_o. p* maps the "periphery" $\widetilde{K}$ of K onto the periphery $\widetilde{K}_o$ of a polygon K_o on S_o. Let $A_1 A_2 m... A$ be all the knots of $\widetilde{K}$ which are equivalent to one of them, the image of which is A_o on $\widetilde{K}_o$. ν is called the characteristic of K. Let Ω be the center of K; this corresponds to a point Ω_0 in K_0 (Ω_0 is called the center of K_0). Let us draw in K_0 the "radius" $\Omega_0 A_0$ and in K the radii ΩA_i $(i=1,2,\ldots,\nu)$. Let us establish a correspondence between the lines ΩA_i and $\Omega_0 A_0$ such that Ω_0 and A_0 are the images of Ω and A_i. Thus there is a topological mapping $\check{p}$ of the boundary of the sector $A_i \Omega A_{i+1}$ onto that of the sector $A_0 \Omega_0 A_0$ with angle 2π. The restriction of $\check{p}$ to T is p^*. $\check{p}$ can be extended as a topological mapping of the interior of the sector $A_i \Omega A_{i+1}$ and the interior of the sector $A_0 \Omega_0 A_0$. Then p^* extends over K, as a mapping p of K onto K_0 such that (K,p) is a covering of K_0, with multiplicity ν; (K,p) is ramified only at the center of K_0, and there with multiplicity ν.

The extension p of p^* can be carried out in every polygon of S. (S,p) is a covering of S_0 which is ramified at the centers Ω_0^i of some of the polygons described on S_0 by T_0. We shall show that this covering is regular. Let us look at a polygon K_0 of S_0. One can characterize K_0 on S_0 by giving one of its knots, for instance N_0, and the types of the two sides which end at N_0. All the polygons of S which are mapped onto K_0, by p, have among their vertices, at least one knot of the set $p^{-1}(N_0)$. All the knots of this set are equivalent, hence all the polygons with one of these knots among their vertices, which are bounded by sides of the same type as these of K_0, have the same characteristic.

THEOREM 4. With any surface Σ, which can be represented by a regularly ramified topological tree, without infinite polygons, one can correlate a closed surface Σ_0 and a mapping π of Σ onto Σ_0 such that (Σ,π) is a covering of Σ_0 which is regularly ramified at a finite number of points (see def. 3, A-1).

In fact $\Sigma \xrightarrow{\sigma} S$ (where σ is one-to-one and continuous). If (S,p) is a covering of S_0 which is regularly ramified, then $T_0 = p(T)$ represents a surface Σ_0 (with the help of an auxiliary mapping of Σ_0 onto B_0 in order to get a covering of B_0) Σ_0 is a closed surface and $\Sigma_0 \xrightarrow{\sigma_0} S_0$. Then there exists a mapping $\pi = \sigma_0^{-1} \cdot p \cdot \sigma$ of Σ onto Σ_0 such that (Σ,π) is a regularly ramified covering of Σ_0.

§4. T SIMPLY CONNECTED

Suppose now that T is regularly ramified and simply connected, i.e., T is traced on the plane S. We have then the following

THEOREM 5. [16]. Given on a closed surface S_o n points Ω_o^i $(i=1,2,\ldots,n)$ and n integers ν_i, each associated with one of the Ω_o^i, then there exists a closed surface S_1 and a mapping p_o of S_1 onto S_o such that (S_1, p_o) is a covering of S_o which is regularly ramified with multiplicity ν_i at the points Ω_o^i and is ramified only at this point (except if $n=2$ with $\nu_1 \neq \nu_2$ and S_o of genus zero).

Let us choose a particular branch $\hat{p}_o^{-1}$ of p_o^{-1} and let us show that $p_1 = \hat{p}_o^{-1}$ p is single-valued.

(i) Locally. Let $A \in S$. If $A \neq \Omega_1$, p_2 is locally one-to-one like p and $\hat{p}_o^{-1}$. If $A \equiv \Omega_1$, let $A_o = p(A)$, $A_1 = \hat{p}_o^{-1}(A_o)$. Let us choose on S_o a neighborhood V_o of A_o which is identical with or included in the polygon of S_o which contains A_1. Let V and V_1 be the images of V_o by p^{-1} and $\hat{p}_o^{-1}$. p_1 is a one-to-one mapping of V onto V_1.

(ii) In the large. Let C be a closed curve in the plane S and A a point on C. If A is not a knot of T, let us replace A by a vertex A' of the polygon Δ containing A, and the part of C lying in Δ by another curve lying in the same polygon (or its boundary) but passing through A'. Then by (i) the new curve is closed or open according as C was open or closed. So, we may assume that A is a knot of T.

Let A_1 and A_1' be the ends of C_1. Let us denote by (r) the following operation: if γ is an arc of C in a polygon K or on its boundary (we suppose K is such that its intersection with the interior of C is not empty), then let us replace γ by the part of the boundary of K which is inside C. The operation (r) does not change the points A_1 and A_1'.

By performing the operation (r) again a finite number of times, one can reduce C to the point A. Then $A_1 = A_1'$ and C_1 is a closed curve.

So, (S, p_1) is a relatively unramified covering of S_1. Let $T_1 = p_1(T) = \hat{p}_o^{-1}(T_o)$ and (Σ_1, τ_1) be the covering of B_o which is represented by $T_1 \cdot \Sigma_1 \xrightarrow{\sigma_1} S_1$ where σ_1 is continuous and one-to-one. Let $\pi_1 = \sigma_1^{-1} \circ p_1 \circ \sigma$. (Σ_1, π_1) is a relatively unramified covering of Σ_1. If we notice that Σ_1 is a closed surface we get the following result.

THEOREM 6. Any simply connected surface Σ which is represented by a regularly ramified topological tree, without infinite polygons, can be mapped on a closed surface Σ_1, by π_1, such that (Σ, π_1) is a relatively unramified covering of Σ.

§5. RIEMANN SURFACES

If B_0 and Σ are Riemann surfaces and (Σ, τ) a Riemannian covering of B_0, then τ_0 and τ are Riemannian projections, as $\pi = \tau_0^{-1} \circ \tau$, $\pi_0 = \tau_0^{-1} \circ \tau_1$ and $\pi_1 = \tau_1^{-1} \circ \tau = \pi_0^{-1} \circ \pi$ since π, π_0, π_1 which are defined by these relations are single-valued.

THEOREM 7. Any simply connected Riemann surface, which is represented by a regularly ramified topological tree, without infinite polygons, can be mapped by a Riemannian projection ω onto a closed Riemann surface, such that $(\mathcal{R}, \omega)$ is a relatively unramified covering of $\mathcal{R}_1$.

Let us notice that any closed Riemann surface $\mathcal{R}_1$ can be mapped (by a Riemannian projection) on a closed Riemann surface of genus zero, so that $\mathcal{R}_1$ can be represented by a regularly ramified topological tree without infinite polygons.

The relationship between the surfaces are shown in the following picture where the signs $\leftrightarrow$ denote topological mappings (one-to-one), $\rightarrow$ continuous mapping, $\Rightarrow$ Riemannian projection.

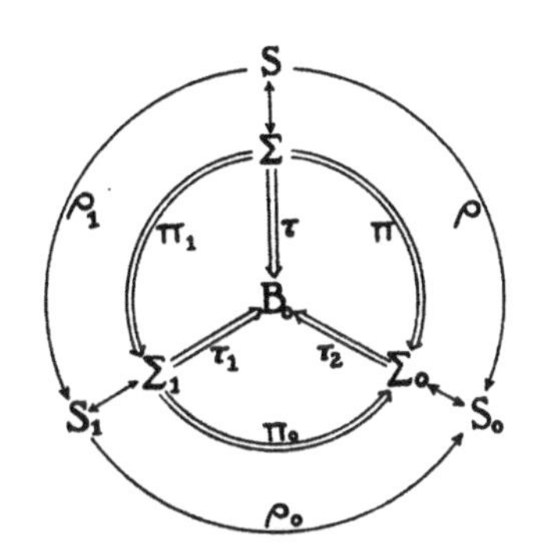

D. AUTOMORPHISM FUNCTIONS AND TOPOLOGICAL TREES

§1. EQUIVALENT KNOTS

Let $(\mathcal{R}, \omega)$ be a Riemannian covering of a Riemann surface B_0 which is conformally equivalent to the plane minus a finite number of points. Every knot of T represents one half of a fundamental domain of $\mathcal{R}$ (with respect to ω). The correlation of two knots of the same kind on T defines an automorphism function ω, whose multiplicity and branch points, we can study on T itself.

Suppose that ω has an automorphism function which is one-to-one. Let A_1 and A_2 be two knots of T, which are the images of two half-domains of $\mathcal{R}$, correlated by q. To any closed curve on T, with origin in A_1, there corresponds a closed curve with its origin in A_2 and conversely. Then A_1 and A_2 are equivalent. Conversely if T has two equivalent knots, the autormorphism function ω defined by the correlation between these two knots is one-to-one. If one considers the global uniformizing function f of $\mathcal{R}$, then $\omega \cdot f^{-1}$ is a single-valued function in the plane, or in the circle, automorphic with respect to a linear function which does not belong to the group obtained by the covering of $\mathcal{R}$. If $\omega \cdot f^{-1}$ is totally automorphic, the cell of T has only two knots of different kinds and conversely.

§2. APPLICATION

We will show, by an example, that a meromorphic function in the unit circle can have an automorphism function which is single-valued, but whose inverse is multi-valued (this is not true for meromorphic functions in the whole plane, by a theorem of Shimizu).

Let us take as the surface B_0, the sphere on which we fix six points. Let us build the covering of B_0 which can be represented by the following topological tree:

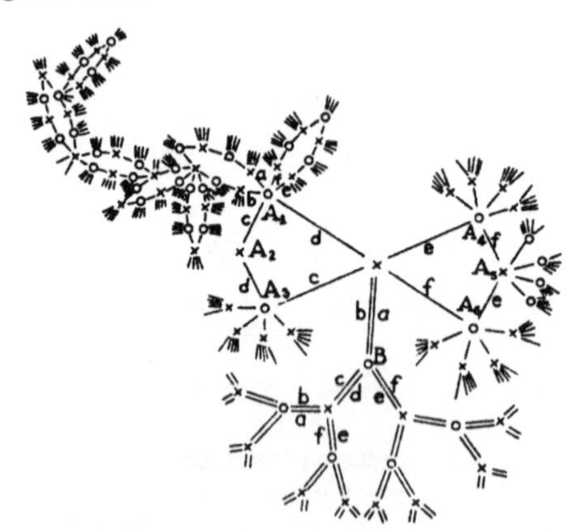

Roughly speaking, this covering is a Riemann surface where the branch points lie over the six given points on the sphere B_0 which is already itself a Riemann surface.

On this tree each knot "above" all the A's, i.e. in the upper part, belong to three octogons, whereas each knot "below" B belongs to three flat polygons (with only two sides). Every automorphism function of the global uniformizing function of $\mathcal{R}$, which correlates a knot below B to a knot above all A's is single-valued but not its inverse.

By Shimizu's theorem, $\mathcal{R}$ is necessarily hyperbolic. This can also be deduced from Wittich's criterion.

BIBLIOGRAPHY

[1] WEYL, H., Die Idee der Riemannschen Fläche, 1913.

[2] FATOU, P., Sur les équations fonctionnelles (Bull. Soc. Math. France 67-68 (1919-1920).

[3] DE KEREKJARTO, Vorlesungen über Topologie, Springer 1923.

[4] SHIMIZU, T., On the fundamental domains and the groups for meromorphic functions, Jap. Journ. Math. 8 (1931).

[5] SEIFERT and THRELFALL, Lehrbuch der Topologie, Teubner 1936.

[6] AHLFORS, L., Zur Theorie der Überlagerungsflachen, Acta Math. 65 (1935).

[7] AHLFORS, L., Geometrie der Riemannschen Flachen, Congrés internation-al des mathématiciens, Oslo, 1936.

[8] STOÏLOW, Lecons sur les principes topologiques de la théorie des fonctions analytiques, Gauthiers-Villars 1938.

[9] MARTY, F., Sur les groupes et hypergroupes attachés à une fraction rationelle Ann. Ec. Norm. Sup. 53 (1936).

[10] NEVANLINNA, R., Eindeutige Analytische Funktionen, Springer (1936).

[11] HALLSTROM, GUNNAR af., Über Substitutionen, die eine rationale Funktion invariant lassen, Acta. Acad. Aboens 15; 6 (1946).

[12] HÄLLSTRÖM, GUNNAR af., On the study of algebraic functions of automorphism by help of graphs, X skand Math. Kongr. København (1946).

[13] HÄLLSTRÖM, GUNNAR, af., Über die automorphiefunktionen meromorpher Funktionen, Acta. Acad. Aboens 16; 4 (1946).

[14] CHEVALLEY, C., Theory of Lie Groups, Princeton 1942.

[15] FOURÈS, L., Sur la théorie des surfaces de Riemann, Ann. Ec. Norm. Sup. 68 (1951).

[16] FOURÈS, L., Sur les recouvrements régulièrement ramifiés, Bull. Sci. Math. 76 (1952).

[17] FOURÈS, L., Le problème des translations isothermes, Ann. Inst. Fourier, Grenoble, (1952).

PARTIAL DIFFERENTIAL EQUATIONS AND PSEUDO-ANALYTIC FUNCTIONS ON RIEMANN SURFACES[1]

Lipman Bers

Riemann surfaces were introduced by Riemann as a tool in the investigation of multiple-valued analytic functions. The ideas and methods of Riemann's function theory can also be used in studying multiple-valued solutions of linear partial differential equations of elliptic type. We intend to outline in the sequel how this may be done (omitting all proofs and details).

§1. PARTIAL DIFFERENTIAL EQUATIONS AND PSEUDO-ANALYTIC FUNCTIONS[2]

Our first task is to write the general linear partial differential equation of second order and elliptic type in a conformally invariant form. Every such equation can be reduced, at least locally[3], to the form

$$\phi_{xx} + \phi_{yy} + \alpha(x,y)\phi_x + \beta(x,y)\phi_y = 0. \tag{1}$$

We assume that α and β are defined and satisfy a Hölder condition in a

---

[1] Some of the results presented in this paper have been obtained during work on a project sponsored by the OOR.

[2] The theory of pseudo-analytic functions in plane domains was announced in L. Bers, Proc. Nat. Acad. Sci., vol. 36, 1950, pp. 130-136; vol. 37, 1951, pp. 42-47. (cf. also the mimeographed lecture notes "Theory of pseudo-analytic functions," New York University, to appear). It contains as special cases previous generalizations of complex function theory due to E. Beltrami (Opere mat., vol. 3, Milano, 1911, pp. 115-128, 349-377), L. Bers and A. Gelbart (Quart. Appl. Math., vol. 1, 1943, pp. 168-188; Trans. A.M.S., vol. 56, 1944, pp. 67-93; Ann. Math., vol. 48, 1947, pp. 342-357; cf. also L. Bers, Amer. J. Math., vol. 72, 1950, pp. 705-712), M. A. Lukomskaya (Doklady Akad. Nauk, vol. 73, 1950, pp. 888-895), A. I. Markushevitch (as reported by I. G. Petrovskii, Uspekhi Mat. Nauk, vol. 1, 1946, pp. 44-70), G. N. Položii (Doklady Akad. Nauk, vol. 58, 1947, pp. 452-498; vol. 60, 1948, pp. 769-772). The possibility of such a theory has been foreseen by Picard (C. R., vol. 112, 1891, pp. 168-188). Results on pseudo-analytic functions on Riemann surfaces are presented here for the first time. Multiple-valued solutions of linear elliptic partial differential equations have been considered, in an unsystematic way, by many authors. The analogy with the theory of Abelian integrals, however, has been pursued, as far as I know, only by S. Bergman in a paper dealing with solutions of self-adjoint equations in a multiply connected domain (Duke Math. J. vol. 14, 1947, pp. 349-366).

[3] More precisely: in every domain in which there exists a positive solution.

bounded domain D_o. Then there exist in D_o functions $\sigma(x,y) > 0$ and $\tau(x,y)$ such that equation (1) is equivalent to the system

$$\phi_x = \tau\psi_x + \sigma\psi_y ,$$
(2)
$$\phi_y = -\sigma\psi_x + \tau\psi_y .$$

The functions σ, τ have Hölder-continuous partial derivatives (or, as we shall say, belong to the class H^1) in D_o.

Now set $z = x + iy$, and let[4] $F(z), G(z)$ be two fixed complex-valued functions of class H^1 defined in D_o, such that $\mathrm{Im}(\overline{F}G) > 0$. Then every function $w(z)$, $z \in D \subset D_o$, admits the unique representation

$$w(z) = \phi(z)F(z) + \psi(z)G(z) \tag{3}$$

with real ϕ, ψ. The (F,G)-derivative of $w(z)$ at $z = z_o$ is defined as

$$\dot{w}(z_o) = \lim_{z \to z_o} \frac{w(z) - \phi(z_o)F(z) - \psi(z_o)G(z)}{z - z_o} , \tag{4}$$

provided that this limit exists and is finite. The function $w(z)$ is called pseudo-analytic (of the first kind) in D with respect to the "generating pair" (F,G) if $\dot{w}(z)$ exists at all points of D; the function

$$\omega(z) = \phi(z) + i\,\psi(z) \tag{5}$$

is then called pseudo-analytic of the second kind. It turns out that the existence of $\dot{w}(z)$ at all points of D implies not only the existence but also the continuity of the derivatives w_x, w_y, and the equations

$$F\phi_{\overline{z}} + G\psi_{\overline{z}} = 0, \tag{6}$$

$$\dot{w} = F\phi_z + G\psi_z , \tag{7}$$

where

$$\frac{\partial}{\partial z} = \frac{1}{2}\left(\frac{\partial}{\partial x} - i\frac{\partial}{\partial y}\right), \quad \frac{\partial}{\partial \overline{z}} = \frac{1}{2}\left(\frac{\partial}{\partial x} + i\frac{\partial}{\partial y}\right) .$$

Conversely, if w_x and w_y exist and are continuous, equation (6) guarantees the pseudo-analyticity of the functions (3), (5).

Now choose (F,G) so that $(G/F) = -\tau + i\sigma$. Then equation (6) becomes identical with (2) and the theory of (F,G)-pseudo-analytic functions with that of solutions of (1). For $(F,G) = (1,i)$ we obtain classical function theory.

We observe that the existence of the limit (4) is a conformally invariant condition.

[4] We write functions of x and y as functions of z without implying analytic dependence.

§2. SOME PROPERTIES OF PSEUDO-ANALYTIC FUNCTIONS

We call two functions, $f(z)$ and $w(z)$, defined in a domain D similar if

$$f(z) = e^{s(z)}w(z),$$

where $s(z)$ is continuous on the closure $\overline{D}$ of D. A basic property of pseudo-analytic functions is expressed by the theorem (<u>similarity principle</u>) which asserts that every pseudo-analytic function w (of the first kind) is similar to an analytic function f, and every analytic function $f(z)$, $z \in D \subset D_0$, is similar to some pseudo-analytic function. This theorem implies, among other things, that pseudo-analytic functions possess the unique continuation property[5], and that we may talk about zeros and poles of definite orders. Another fundamental theorem states that pseudo-analytic functions of the second kind are <u>interior transformations</u>, that is, of the form $\omega(z) = g[\chi(z)]$, where $\chi(z)$ is a homeomorphism and $g(z)$ an analytic function.

Together with the generating pair (F,G) we consider the generating pair (F^*,G^*) defined by

$$F^* = \frac{2\overline{G}}{F\overline{G} - \overline{F}G}, \quad G^* = \frac{2\overline{F}}{F\overline{G} - \overline{F}G}, \tag{8}$$

so that

$$F^{**} = F, \quad G^{**} = G.$$

From (3), (5), (7) it follows that if $w(z)$ and $\omega(z)$ are corresponding pseudo-analytic functions of the first and second kinds, then

$$\omega(z) = \omega(z_0) + \operatorname{Re} \int_{z_0}^{z} F^*\dot{w}dz - i \operatorname{Re} \int_{z_0}^{z} G^*\dot{w}dz.$$

Conversely, if $W(z)$ is any continuous function defined in a simply connected domain $D \subset D_0$, and if

$$\operatorname{Re} \oint_C F^*Wdz - i \operatorname{Re} \oint_C G^*Wdz = 0$$

for every closed curve C in D, then there exists in D a pseudo-analytic function $w(z)$ such that $\dot{w}(z) = W(z)$.

On the other hand we can find in every domain $D_1, \overline{D}_1 \subset D_0$, a generating pair (F_1,G_1) called a <u>successor</u> of (F,G), such that the

---

[5] This also follows from a theorem of Carleman (C. R. vol. 297, 1933, pp. 471-474).

(F,G)-derivatives of (F,G)-pseudo-analytic functions are (F_1,G_1)-pseudo-analytic. Indeed, we can embed (F,G) in a generating sequence, that is we can find in D_1 a sequence of generating pairs $\{(F_\nu,G_\nu)\}$ such that $(F_0,G_0) = (F,G)$ and $(F_{\nu+1},G_{\nu+1})$ is a successor of (F_ν,G_ν), $\nu = 0, \pm 1, \pm 2, \ldots$.

§3. PSEUDO-ANALYTIC FUNCTIONS AND DIFFERENTIALS ON RIEMANN SURFACES

Let S be a Riemann surface and F,G two functions defined on S such that $\operatorname{Im}(\overline{F},G) > 0$. (F,G) is called a generating pair if in the neighborhood of every point p of S, and for every local parameter $z = x + iy$ belonging to p, F and G considered as functions of (x,y) are of class H^1. Every function w defined on S admits the unique representation $w = \phi F + \psi G$ with real ϕ, ψ. w is called (F,G)-pseudo-analytic of the first kind if it is so when considered as a function of any local parameter; $\omega = \phi + i\psi$ is then an (F,G)-pseudo-analytic function of the second kind.

Let dw be a continuous differential on S, and C a closed continuously differentiable curve. The number

$$\operatorname{Re} \oint_C F^* \, dw - i \operatorname{Re} \oint_C G^* \, dw$$

is called the (F,G)-period of dw over C. If it vanishes whenever C is homologous to 0, dw is called (F,G)-pseudo-analytic. By an (F,G)-pseudo-analytic meromorphic function on S (henceforth referred to as an (F,G)-function) we mean a function w which is (F,G)-pseudo-analytic on $S - \Sigma$, Σ being a discrete set on S, and having poles at all points of Σ. An (F,G)-pseudo-analytic meromorphic differential on S ((F,G)-differential) is defined similarly. Note that since the similarity principle may be applied in the neighborhood of every point on S, the poles and zeros of (F,G)-functions and (F,G)-differentials are of definite multiplicities.

The residue of an (F,G)-differential dw at a point p of S where dw has a pole is defined as the (F,G)-period of dw on a simple closed curve around and sufficiently close to p. An (F,G)-function w such that

$$dw = \dot{w}(z)dz,$$

$\dot{w}(z)$ being computed in terms of the local parameter z, exists if and only if all residues of dw and all (F,G)-periods of dw on non-bounding closed curves vanish. If this condition is violated, then

$$\omega(q) = \operatorname{Re} \int_{q_0}^{q} F^* dw - i \operatorname{Re} \int_{q_0}^{q} G^* dw$$

is a multiple-valued (F,G)-function of the second kind (Abelian (F,G)-integral).

§4. EXISTENCE THEOREM FOR CLOSED SURFACES AND THE RIEMANN ROCH THEOREM

Let S be a closed Riemann surface of genus g. While the (F,G)-functions on S do not form a field but only a vector space over the reals, it is still convenient to use the language of divisors. A divisor $d = p_1^{\alpha_1} \dots p_r^{\alpha_r}$ is an element of the multiplicatively written free commutative group generated by the points p of S; the sum $\alpha_1 + \dots + \alpha_r$ is the order of d; d is integral if $\alpha_1 \geq 0, \dots, \alpha_r \geq 0$. If an (F,G)-function or differential has zeros at $p_1, \dots, p_r$ of multiplicities $\alpha_1, \dots, \alpha_r$ and poles at $q_1, \dots, q_s$ of multiplicities $\beta_1, \dots, \beta_s$, and no other zeros or poles, then it is "represented" by the divisor $d = p_1^{\alpha_1} \dots p_r^{\alpha_r} q_1^{-\beta_1} \dots q_s^{-\beta_s}$, and is a "multiple" of every divisor d_1 such that dd_1^{-1} is integral.

From the interior character of pseudo-analytic functions it follows at once that every (F,G)-function is represented by a divisor of order 0, every (F,G)-differential by one of order $2(g - 1)$, that the sum of all residues of an (F,G)-differential dw is 0, and that $dw \equiv 0$ if it has no poles and purely imaginary periods.

On the other hand, the similarity principle can not be applied to S as a whole, so that the existence of (F,G)-differentials and functions must be established directly. The corresponding classical theorems can be proved either by a variant of Dirichlet's principle or by one of the alternating method. Since we consider equations which do not necessarily come from a variational problem, only the latter path is open. It leads to the following

> THEOREM 1. Let $p_1, \dots, p_k$, $k \geq 0$, be distinct points on S and let dw_j, $j = 1,2,\dots,k$, be an (F,G)-pseudo-analytic meromorphic differential defined in a neighborhood of p_j and having at p_j a pole with residue σ_j. If $\sigma_1 + \dots + \sigma_k = 0$, then there exists on S exactly one (F,G)-differential dw such that $dw - dw_j$ is regular at p_j $(j = 1,2,\dots,k)$ and all periods of dw on non-bounding curves are purely imaginary.

We observe that the differentials dw_j can be constructed by the similarity principle. For $k = 0$ we obtain the existence of 2g linearly independent everywhere regular (F,G)-differentials.

The pseudo-analytic meromorphic differentials on S are connected by reciprocity relations, analogous to the well-known relations between Abelian differentials. These relations, which we shall not write down explicitly, always involve an (F,G)-differential and an (F^*,G^*)-differential. They lead to the following

THEOREM 2. (Generalized Riemann-Roch Theorem). Let d be a divisor on S of order m, A the number of linearly independent (F^*,G^*)-functions which are multiples of $1/d$, B the number of linearly independent (F,G)-differentials which are multiples of d. Then $A - B = 2(m + 1 - g)$.

Set $d = p_1^{-\alpha_1} \dots p_k^{-\alpha_k}$, $\alpha_1 > 0, \dots, \alpha_k > 0$ and $n = -m = \alpha_1 + \dots + \alpha_k \geq k + 2$. Then $A = 0$ and $B = 2(n + g - 1) > 0$. Let dw be an (F,G)-differential which is a multiple of d. It will be the differential of an (F,G)-function if its residues and its periods vanish, that is if $2(k + g)$ linear homogeneous conditions are satisfied. Hence (F,G)-functions exist on S. The same result can be obtained by noting that $B = 0$ for $m > 2(g - 1)$.

REMARK. A universally accepted heuristic principle states that two mathematical theories exhibiting similar theorems are special cases of a more general theory. The classical Riemann-Roch theorem, which can be proved either purely algebraically or by potential-theoretical arguments, illustrates a "dual" principle: if a central theorem in a theory can be proved in two essentially different ways, the theory is an "intersection" of two distinct generalizations. Indeed, the theory of algebraic functions of a complex variable may be considered either as a special case of the theory of algebraic functions over a general number field, or as a special case of the theory of linear partial differential equations.

§5. THE CASE $g = 0$

In this case we may represent S as the full function-theoretical plane. We deal then with a generating pair $(F(z),G(z))$ defined for all z and such that $(\alpha):(F(1/z),G(1/z))$ is a generating pair for $|z| < \infty$. There is no need to use the general existence theorems of the preceding section, since the similarity principle can be applied to S.

Using this principle we can show that for every $a \neq 0$, z_0 and integer n there exists one and only one (F,G)-function $w(z) = Z^{(n)}(a,z_0;z)$ such that $(z - z_0)^{-n}w(z)$ is uniformly bounded, bounded away from zero, and has the limit a at $z = z_0$. For $a = 0$ we set $Z^{(n)}(0,z_0;z) = 0$. Every (F,G)-function on S is a sum of "formal powers" $Z^{(n)}$.

Using the formal powers one obtains, for (F,G)-pseudo-analytic functions defined in subdomains of the z-plane, analogues of the Cauchy formula,[6] the Runge approximation theorem and of the Taylor and Laurent expansions.[7]

A generating pair satisfying condition (α) will not, in general, have a successor satisfying the same condition. The preceding results, however, remain valid if (α) is replaced by the less stringent condition (β): there exist constants $R_0 > 0$, $K > 0$, $0 < \alpha < 1$, $\sigma > 1$ such that in every domain $|z| > R < R_0$ the derivatives of F and G are bounded in modulus by $KR^{-\sigma}$ and satisfy a Hölder condition with exponent α and coefficient $KR^{-\sigma}$. Assuming this condition set $F_1(z) = \dot{Z}^{(1)}(1,0;z)$, $G_1(z) = \dot{Z}^{(1)}(i,0;z)$. It can be shown that (F_1,G_1) is a generating pair satisfying (β), and a successor of (F,G). It is easy to see that (F^*,G^*) satisfies (β), and that the "star" of a successor of (F^*,F^*) is a predecessor of (F,G). Thus we obtain the result: a generating pair (F,G) satisfying (β) can be embedded in a generating sequence $\{(F_\nu,G_\nu)\}$, such that all (F_ν,G_ν) satisfy (β). The pairs (F_ν,G_ν) are determined uniquely, except for obvious trivial modifications.

Denote the (F_ν,G_ν)-pseudo-analytic formal powers by $Z_\nu^{(n)}(a,z_0;z)$. Then every (F,G)-differential on S is a sum of differentials $Z_1^{(n)}(a,z_0;z)dz$. In particular, $Z_1^{(-1)}(a,z_0;z)dz$ has non-vanishing residues at $z = z_0$ and $z = \infty$, provided $a \neq 0$. Using this differential one can obtain the fundamental solution of the partial differential equation (1).

§6. THE CASE g = 1

Let (F,G) be a doubly periodic generating pair. We mean by this that $F(z)$ and $G(z)$ are defined for all values of z and possess two linearly independent periods 2ω, $2\omega'$. The investigation of pseudo-analytic functions on a surface S of genus 1 amounts to the study of (F,G)-pseudo-analytic functions having the same periods. Using the general theorems stated above essential parts of the theory of elliptic functions can be duplicated.

---

[6] For other generalizations of Cuachy's formula to differential equations cf. S. Bergman, Trans. A.M.S., vol. 63, 1947, pp. 432-497, G. N. Položii, Mat. Sbornik, vol. 24 (66), 1949, pp. 375-384, B. V. Šabat, Doklady Akad. Nauk, vol. 69, 1949, pp. 305-308.

[7] The result concerning formal power series stated in the two notes quoted in Footnote 1 is incomplete. The final result has been obtained jointly by S. Agmon and the author, Proc. A.M.S., vol. 3, 1952, pp. 757-764.

In particular, there will exist two linearly independent everywhere regular (F,G)-pseudo-analytic functions, $W_1(z)$ and $W_2(z)$, such that

$$W_j(z + 2m\omega + 2n\omega') = W_j(z) + (m\xi_{j,1} + n\xi_{j,2})\, F(z)$$
$$+ (m\eta_{j,1} + n\eta_{j,2})\, G(z),$$

$m,n = 0, \pm 1, \pm 2, \ldots,$ $j = 1,2$ where, ξ_{j1}, η_{j1} are real constants. Set $F_1(z) = \dot{W}_1(z)$, $G_1(z) = \dot{W}_2(z)$. Since $\dot{F}(z) = \dot{G}(z) = 0$, F_1 and G_1 have periods 2ω, $2\omega'$. Since an everywhere regular (F,G)-differential on S has no zeros, F_1 and G_1 never vanish, and for the same reason $\operatorname{Im}(\overline{F}G) > 0$. It can be shown that (F_1,G_1) is a successor of (F,G), and since (F^*,G^*) has the same periods as (F,G) we obtain the result: every doubly-periodic generating pair (F,G) can be embedded in a generating sequence $\{(F_\nu,G_\nu)\}$ such that all F_ν,G_ν have the same periods.

§7. AN APPLICATION

In order to show the application of the general theorems by at least one example, consider the equation (1) with $\alpha(x,y)$, $\beta(x,y)$ defined and Hölder-continuous for all x,y. We also assume (essentially for the sake of simplicity) that α and β vanish outside of a large circle $|z| = R$. Consider the following problem: to find all m-valued solutions of (1) which are defined for all values of (x,y), except for a finite number of excluded points $z_1, z_2, \ldots, z_k, \infty$ and such that for a sufficiently large $n > 0$, $\phi(x,y) = O(|z - z_k|^{-n})$, $z \to z_j$, $j = 1,2,\ldots,k$, $\phi(x,y) = O(|z|^n)$, $z \to \infty$.

To solve this problem let S be any unbounded m-sheeted covering surface of the full function-theoretical plane with a finite number of branch-points located over $z_1,\ldots,z_k, \infty$. In an obvious way S may be made into a Riemann surface. We can find a generating pair $(\widetilde{F},\widetilde{G})$ defined on the full function-theoretical plane such that solutions of (1) are the real parts of $(\widetilde{F},\widetilde{G})$-pseudo-analytic functions of the second kind. Now set, for every point p of S located over z, $F(p) = \widetilde{F}(z)$, $G(p) = \widetilde{G}(z)$. It is easy to check that (F,G) is a generating pair on S. The general theorem stated above assures the existence of an (F,G)-differential dw on S with purely imaginary residues and periods. Set

$$\phi(q) = \operatorname{Re} \int_{q_0}^{q} F^* \, dw.$$

Then ϕ is a single-valued function on S, and considered as a function of (x,y) it is a solution of our problem. It is not difficult to see that all solutions can be obtained in this way.

§8. OPEN QUESTIONS

The theory of pseudo-analytic functions and differentials on <u>open surfaces</u> is certainly a desideratum. This theory will present a novel feature: the behaviour of functions and differentials on an open Riemann surface must depend not only on the conformal structure of the surface but also on the behaviour of the generators $F(p)$, $G(p)$ as p tends toward the ideal boundary. This is closely connected with the as yet rather incomplete theory of equations of mixed type.

As far as multiple-valued and singular solutions of non-linear equations are concerned, only the special though probably typical case of the equation of minimal surfaces has been explored thoroughly.[8] Riemann surfaces are here, too, an important tool, but the results are quite unlike those encountered in the linear case.

We mention finally that the work presented here may be thought of as a generalization of the simplest case of the theory of p-dimensional harmonic forms on n-dimensional Riemannian manifolds: the case $p = 1$, $n = 2$. It is likely that similar generalizations are possible for all even n and $p = n/2$.

---

[8] Y. W. Chen, Ann. Math., vol. 49, 1948, pp. 790-806, L. Bers, Ann. Math., vol. 53, 1951, pp. 364-486; Journal d'Analyse Math. (Jerusalem), vol. 1, 1951, pp. 43-58. Recently R. S. Finn obtained extensions of some of these results to more general non-linear equations.

DIRICHLET'S PRINCIPLE AND SOME INEQUALITIES IN THE THEORY OF CONFORMAL MAPPING

Zeev Nehari

Many of the inequalities of function theory and potential theory may be reduced to statements regarding the properties of harmonic domain functions with vanishing boundary values, that is, functions which can be obtained from the Green's function by means of elementary processes. To derive these inequalities, a large number of different techniques and procedures have been used. It is the aim of this paper to show that many of the known inequalities of this type, and also others which seem to be new, can be obtained as a simple consequence of the classical minimum property of the Dirichlet integral. In addition to the resulting simplification, this method has the further advantage of being capable of generalization to a wide class of linear partial differential equations of elliptic type in two or more variables. Because of the limited space available, we shall here confine ourselves to a small number of examples, and some of the proofs will only be sketched. A fuller account will appear elsewhere.

The domains we shall consider will be assumed to be bounded by a finite number of analytic curves and they will be embedded in a given Riemann surface R of finite genus. The symbol $S(z)$ will be used to denote a "singularity function" with the following properties: $S(z)$ is real, harmonic and single-valued on R, with the exception of a finite number of points at which $S(z)$ has specified singularities. $S(z)$ is thus the real part of a properly normalized Abelian integral.

We now prove the following statement.

THEOREM 1. Let D and D^* be two domains embedded in R such that $D \subset D^*$ and that $D^* - D$ contains no singularities of $S(z)$, and let C and C^* denote the boundaries of D and D^*, respectively. If the function $p(z)$ vanishes on C and is such that $h(z) = p(z) + S(z)$ is harmonic and single-valued in D, and if $p^*(z)$ and $h^*(z)$ denote the corresponding functions associated with D^*, then

$$\int_C h(z) \frac{\partial p(z)}{\partial n} ds \geq \int_{C^*} h^*(z) \frac{\partial p^*(z)}{\partial n} ds, \tag{1}$$

where the differentiation is performed with respect to the outer normal.

By the minimum property of the Dirichlet integral, we have

$$\iint_{D^*} (h_x^{*2} + h_y^{*2})\, dxdy \leq \iint_{D^*} (u_x^2 + u_y^2)\, dxdy, \tag{2}$$

where $u(z)$ is continuous in the closure of D^* and has on C^* the same boundary values as $h^*(z)$; the derivatives u_x and u_y need only be continuous in subdomains of D^* into which D^* is divided by a finite number of smooth arcs or curves. In view of $h(z) = p(z) + S(z)$ and $h^*(z) = p^*(z) + S(z)$, and the fact that $p(z)$ and $p^*(z)$ vanish on C and C^*, respectively, the function $u(z)$ in (2) may be defined as follows:

$$u(z) = h(z), \quad z \in D$$
$$u(z) = S(z), \quad z \in D^* - D.$$

Inserting this in (2) and using Green's formula, we obtain

$$\int_{C^*} h^* \frac{\partial h^*}{\partial n} ds \leq \int_C h \frac{\partial h}{\partial n} ds + \int_{C^*} S \frac{\partial S}{\partial n} ds - \int_C S \frac{\partial S}{\partial n} ds. \tag{3}$$

In view of $h = p + S$, we have

$$\int_C h \frac{\partial h}{\partial n} ds = \int_C h \frac{\partial p}{\partial n} ds + \int_C S \frac{\partial S}{\partial n} ds,$$

and similarly for h^*. Substituting in (3), we arrive at (1). This proves Theorem 1.

To illustrate the application of (1), take R to be the simple plane and set $S(z) = \sum_{\nu=1}^{n} \alpha_\nu \log |z - \zeta_\nu|$, where the α_ν are arbitrary real constants and $\zeta_1, \ldots, \zeta_n$ are points of a finite domain D. We then have $p(z) = \sum_{\nu=1}^{n} \alpha_\nu\, g(z, \zeta_\nu)$, where $g(z, \zeta)$ is the Green's function of D. If we write $g(z, \zeta) = -\log|z - \zeta| + h(z, \zeta)$, Green's formula yields

$$\int_C h \frac{\partial p}{\partial n} ds = -2\pi \sum_{\nu,\mu=1}^{n} \alpha_\nu \alpha_\mu h(\zeta_\nu, \zeta_\mu).$$

It therefore follows from (1) that the quadratic form

$$Q(\alpha, \alpha) = \sum_{\nu,\mu=1}^{n} \alpha_\nu \alpha_\mu h(\zeta_\nu, \zeta_\mu)$$

grows monotonically with the domain. Since, for a circle of radius ρ about the origin, we have $h(z, \zeta) = \log \zeta + \log|1 - z\zeta^{-2}|$, the value of the form $Q(\alpha, \alpha)$ for this circle will tend to zero if $\alpha_1 + \ldots + \alpha_n = 0$ and $\zeta \to \infty$. It follows that

$$\sum_{\nu,\mu=1}^{n} \alpha_\nu \alpha_\mu h(\zeta_\nu, \zeta_\mu) \leq 0, \quad \left(\sum_{\nu=1}^{n} \alpha_\nu = 0\right), \tag{4}$$

for any finite domain and hence, by approximation, for arbitrary plane domains not containing $z = \infty$. In the case $n = 2$, $\alpha_1 = 1$, $\alpha_2 = 1$, this reduces to

$$h(\zeta,\zeta) + h(\eta,\eta) \leq 2h(\zeta,\eta), \tag{5}$$

an inequality previously obtained by means of Hadamard's variation formula for the Green's function [7,1]. As shown in [1], (5) contains the distortion theorems for schlicht conformal mappings of the unit circle.

Another application of (1) concerns the Bergman kernel function $K(z,\zeta)$ of the domain D [1]. It was shown by Schiffer [6] that

$$K(z, \zeta) = -\frac{2}{\pi} \frac{\partial^2 g(z, \zeta)}{\partial z \partial \bar{\zeta}}, \tag{6}$$

where $g(z, \zeta)$ is the Green's function of D. Schiffer also showed that the function

$$L(z, \zeta) = \frac{2}{\pi} \frac{\partial^2 g(z, \zeta)}{\partial z \partial \zeta} = \frac{1}{\pi(z-\zeta)^2} + \ell(z, \zeta) \tag{7}$$

plays a fundamental role in the theory of conformal mapping. Both $K(z, \zeta)$ and the function $\ell(z, \zeta)$ defined in (7) are regular in D.

If $\alpha_1, \ldots, \alpha_n$ are complex parameters, then

$$p(z) = \mathrm{Re} \left\{ \sum_{\nu=1}^{n} \alpha_\nu \frac{\partial g(z, \zeta_\nu)}{\partial \zeta_\nu} \right\}$$

vanishes on C and may be identified with the function $p(z)$ in (1). Using (1), (6), (7) and Green's formula, we find that the real quantity

$$\sum_{\nu,\mu=1}^{n} \alpha_\nu \bar{\alpha}_\mu K(\zeta_\nu, \zeta_\mu) + \mathrm{Re} \left\{ \sum_{\nu,\mu=1}^{n} \alpha_\nu \alpha_\mu \ell(\zeta_\nu, \zeta_\mu) \right\}$$

decreases if D increases. Some further manipulation shows that

$$\sum_{\nu,\mu=1}^{n} \alpha_\nu \bar{\alpha}_\mu K(\zeta_\nu, \zeta_\mu) \pm \left| \sum_{\nu,\mu=1}^{n} \alpha_\nu \alpha_\mu \ell(\zeta_\nu, \zeta_\mu) \right| \tag{8}$$

decreases with increasing D, a result derived by Bergman and Schiffer [2] by means of a variational method. These authors also show that the inequality

$$\left| \sum_{\nu,\mu=1}^{n} \alpha_\nu \alpha_\mu \ell(\zeta_\nu, \zeta_\mu) \right| \leq \sum_{\nu,\mu=1}^{n} \alpha_\nu \bar{\alpha}_\mu K(\zeta_\nu, \zeta_\mu),$$

which is an immediate consequence of (8), contains Grunsky's necessary and sufficient conditions for the coefficients of a univalent function [5].

We now state a variant of Theorem 1 which is useful in the treatment of function-theoretic extremal problems in multiply-connected domains. For the sake of simplicity, we shall formulate it only for finite plane domains.

THEOREM 2. Let D and D^* be two finite plane domains which are bounded by the closed analytic curves $C_1, \ldots, C_n$ $(C_1 + \ldots + C_n = C)$ and $C_m^*(C_1^* + \ldots + C_m^* = C^*)$, respectively, and let $D \subset D^*$. Let further $p(z)$ be the function which is zero on the outer boundary of D, takes constant values on the other D_ν and is such that $H(z) = S(z) + P(z)$ is harmonic and single-valued and has a single-valued harmonic conjugate in D, where $S(z)$ is the singularity function used in Theorem 1. If $P^*(z)$ and $H^*(z)$ are the corresponding functions defined with respect to D^*, then

$$\int_C H(z) \frac{\partial P(z)}{\partial n} ds \geq \int_{C^*} H^*(z) \frac{\partial P^*(z)}{\partial n} ds. \tag{9}$$

The proof of Theorem 2 is similar to that of Theorem 1, the main difference being that the classical minimum property of the Dirichlet integral is now replaced by the fact that a harmonic function $u(z)$ with a single-valued conjugate minimizes the Dirichlet integral within the class of functions $v(z)$ with the following properties: On C_ν, $u-v = \text{const.}$; v is continuous except at the points of a finite number of closed analytic curves across which $v(z)$ may have a constant jump; v_x and v_y are continuous except across a finite number of analytic arcs. The proof of this minimum property follows by a suitable application of Green's formula if it is observed that the existence of a single-valued conjugate of u is equivalent to the assumption that $\int_\Gamma \frac{\partial u}{\partial n} ds = 0$ for any simple closed curve Γ in D.

As an illustration of how Theorem 2 may be applied, consider the function $F(z,\zeta)$ mapping D conformally onto the unit circle which is furnished with a number of circular slits centered at the origin; $F(z,\zeta)$ will be normalized by the requirement that the unit circumference be the image of the outer boundary and that $F(\zeta,\zeta) = 0$. The function $-\log|F(z,\zeta)|$ is clearly identical with the function $P(z)$ of Theorem 2 if $S(z)$ is taken to be $\log|z-\zeta|$. By Green's formula, we find that in this case

$$\int_C H(z) \frac{\partial P(z)}{\partial n} ds = 2\pi \log |F'(\zeta,\zeta)| .$$

It therefore follows from (9) that $|F'(\zeta,\zeta)|$ decreases monotonically with increasing domain.

If D is the unit circle, then obviously $F(z,0) = z$. Hence, for any multiply-connected domain contained in the unit circle and containing the origin, we have $|F'(0,0)| \geq 1$. This is equivalent to the following well-known result [3,4]:

RESULT. Let $f(z)$ be univalent and $|f(z)| \leq 1$ in D and let $f(\zeta) = 0$ $(\zeta \in D)$. If the outer boundary of D is mapped by $f(z)$ onto the outer boundary of the image domain, then $|f'(\zeta)| \leq |F'(\zeta,\zeta)|$, where $F(z,\zeta)$ is the mapping function defined above.

In a similar way, the extremal properties of other canonical conformal mappings can be derived.

An interesting set of inequalities is obtained if the procedure leading to Theorem 1 is generalized to the case of a domain D which contains a number of disjoint subdomains $D_1,\ldots,D_n$. We now have to introduce a number of singularity functions $S_\nu(z)$, $\nu=1,\ldots,n$, which are harmonic and single-valued in the closure of $D-D_\nu$. Restricting ourselves, for shorter formulation, to the case of plane domains, we then obtain the following result:

THEOREM 3. Let D be a plane domain and let $D_1,\ldots,D_n$ be disjoint subdomains of D. Let $S_\nu(z)$ be the singularity functions defined above and let the functions $p_\nu(z)$ be such that $p_\nu(z) = 0$ on the boundary C_ν of D_ν and that $h_\nu(z) = p_\nu(z) + S_\nu(z)$ is harmonic in D. If the function $P(z)$ is such that $P(z)$ vanishes on the boundary C of D and that $H(z) = P(z) + \sum_{\nu=1}^{n} S_\nu(z)$ is harmonic in D, we have the inequality

$$\sum_{\nu=1}^{n} \int_{C_\nu} h_\nu \frac{\partial p_\nu}{\partial n} ds + \sum_{\nu=1}^{n} \sum_{\mu\neq\nu} \int_{C_\nu} S_\mu \frac{\partial p_\nu}{\partial n} ds \geq \int_C H \frac{\partial P}{\partial n} ds. \quad (10)$$

The proof of Theorem 3 follows from the remark that the function $u(z)$ defined by

$$u(z) = H(z) + \sum_{\mu\neq\nu} S_\mu(z), \qquad z \in D$$

$$u(z) = \sum_{\mu=1}^{n} S_\mu(z), \qquad z \in D - \sum_{\nu=1} D_\nu$$

is continuous in D and has on C the same boundary values as the harmonic function $H(z)$. Using the minimum property of the Dirichlet integral and Green's formula, we arrive, after some manipulation, at (10).

To illustrate Theorem 3, set $S_\nu(z) = \alpha_\nu \log|z - \zeta_\nu|$, where ζ_ν is a point of D and α_ν is a real constant. We then have $p_\nu(z) = \alpha_\nu g_\nu(z, \zeta_\nu)$, where g_ν is the Green's function of D_ν and $P(z) = \sum_{\nu=1}^{n} \alpha_\nu G(z, \zeta_\nu)$, where G is the Green's function of D. In this case, (10) and Green's formula lead to the following result.

RESULT. Let $D_1, \ldots, D_n$ be a number of disjoint domains contained in a domain D and let ζ_ν be a point of D. If $g_\nu(z, \zeta_\nu) = -\log|z - \zeta_\nu| + h_\nu(z, \zeta_\nu)$ and $G(z, \zeta_\nu) = -\log|z - \zeta_\nu| + H(z, \zeta_\nu)$ denote the Green's functions of D_ν and D, respectively, we have the inequality

$$(11) \quad \sum_{\nu=1}^{n} \alpha_\nu^2 h_\nu(\zeta_\nu, \zeta_\nu) + 2 \sum_{\nu \neq \mu} \alpha_\nu \alpha_\mu \log|\zeta_\nu - \zeta_\mu| \leq \sum_{\nu,\mu=1}^{n} \alpha_\nu \alpha_\mu H(\zeta_\nu, \zeta_\mu)$$

where the α_ν are arbitrary real constants.

If we express the various Green's functions in terms of the analytic functions mapping the domains in question onto the unit circle, we can derive from (11) a considerable amount of information concerning these mappings. As a simple example, consider the case $n = 2$, $\alpha_1 = 1$, $\alpha_2 = -1$ and take D to be the unit circle. A simple computation yields the following result.

RESULT. Let $f(z)$ and $g(z)$ be univalent and bounded (by unity) in $|z| < 1$ and let $f(z) \neq g(w)$, where z and w are any two points in the unit circle. Then

$$|f'(z)g'(w)| \leq \left(\frac{1-|f(z)|^2}{1-|z|^2}\right)\left(\frac{1-|g(w)|^2}{1-|w|^2}\right)\left|\frac{f(z)-g(w)}{1-\overline{g(w)}\, f(z)}\right|^2$$

This inequality is sharp and, for two given points $z = z_0$ and $w = w_0$, equality will hold only if the conformal maps yielded by the two functions partition the unit circle along the curve

$$\left|\frac{z - f(z_0)}{1 - \overline{f(z_0)}\, z}\right| = \left|\frac{1 - \overline{g(w_0)}\, z}{z - g(w_0)}\right|$$

If $\sum_{\nu=1}^{n} \alpha_\nu = 0$, we can combine (11) with (4) to obtain the inequality

$$\sum_{\nu=1}^{n} \alpha_\nu^2 h_\nu(\zeta_\nu, \zeta_\nu) + 2 \sum_{\nu \neq \mu} \alpha_\nu \alpha_\mu \log|\zeta_\nu - \zeta_\mu| \leq 0, \quad \sum_{\nu=1}^{n} \alpha_\nu = 0, \tag{12}$$

which holds under the sole assumption that the D_ν be mutually disjoint domains none of which contains the point at infinity. (12) is equivalent to the following result.

RESULT. If the functions $f_1(z), \dots, f_n(z)$ are univalent in $|z| < 1$ and map $|z| < 1$ onto mutually disjoint domains, then

$$\prod_{\nu=1}^{n} |f_\nu'(z_\nu)(1 - |z_\nu|^2)|^{\alpha_\nu^2} \leq \prod_{\nu \neq \mu} |f_\nu(z_\nu) - f_\mu(z_\mu)|^{-2\alpha_\nu \alpha_\mu},$$

where the α_ν are real constants satisfying $\sum_{\nu=1}^{n} \alpha_\nu = 0$, and $z_1, \dots, z_n$ are arbitrary points of the unit circle.

This inequality will be sharp if the system of curves

$$\sum_{\nu=1}^{n} \alpha_\nu \log \left| \frac{1 - z\overline{f(z_\nu)}}{z - f(z_\nu)} \right| = 0$$

divides the plane into precisely n disjoint subdomains.

We finally give a brief indication of a method of proof which can be used for the solution of a certain class of extremal problems and which, in addition to the minimum property of the Dirichlet integral, uses the following simple topological lemma (whose proof we omit).

LEMMA. Let A be a non-selfintersecting analytic arc which connects the points $(0,0)$ and $(1,1)$ in the xy-plane, and denote by B the linear segment connecting these two points. Let $G_1, \dots, G_k$ be the domains whose entire boundaries consist of points of A or B but have otherwise no points in common with either A or B. Then it is always possible to shade some of the domains G_ν in such a way that, except for isolated points, no two domains G_ν which are either both shaded or both unshaded will face each other across B.

Because of the limited space we shall confine ourselves to one simple example illustrating the method in question. Let Δ be the class of finite doubly-connected domains D whose outer boundary is a closed analytic curve D and which do not contain two points α and β within

C, and consider the problem $M(D) = \max.$; $D \in \Delta$, where $M(D)$ is the Riemann modulus of D. If $u(z)$ is the harmonic function in D which has the boundary values 1 and 0 on C and on the inner boundary of D, respectively, then

$$(u,u)_D = \iint_D (u_x^2 + u_y^2)\, dxdy = \frac{2\pi}{\log M(D)}$$

and our problem thus reduces to minimizing the Dirichlet integral in question. We shall show that the problem is solved by a domain D_o whose inner boundary is an analytic slit γ connecting α and β and which is such that the corresponding function $u_o(z)$ satisfies

$$\left(\frac{\partial u_o(z)}{\partial n}\right)_{z=z_1} = \left(\frac{\partial u_o(z)}{\partial n}\right)_{z=z_2} \tag{13}$$

if z_1 and z_2 are two geometrically coinciding points on opposite edges of the slit.

It is clearly enough to show that $M(D) \leq M(D_o)$, where D is a domain whose interior boundary is an analytic curve or an analytic slit. In both cases we can draw an analytic arc δ which connects α and β and has no points in common with D. If we denote the domain bounded by C and δ by D' and set $u(z) = 0$ at the points of D' which do not belong to D, $u(z)$ will be continuous in D' and $(u,u)_D=(u,u)_{D'}$. We now apply the lemma and observe that its purely topological assertion remains true if the linear segment B is replaced by the analytic arc γ and if δ takes the role of the arc A. We denote those domains within C which, by the procedure of the lemma, are left unshaded, by D_1 and the shaded domains by D_2.

We now introduce the function $u_1(z)$ which coincides with $u(z)$ in D_1 and with $-u(z)$ in D_2. Since $u(z)$ vanishes on δ, $u_1(z)$ is continuous in D_o and we obviously have $(u_1,u_1)_{D_o} = (u,u)_D$. Hence

$$(u,u)_D = (u_o,u_o)_{D_o} + 2(u_o,u_1-u_o)_{D_o} + (u_1-u_o,u_1-u_o)_{D_o}. \tag{14}$$

By Green's formula,

$$(u_o,u_1-u_o)_{D_o} = \int_\gamma (u_1-u_o)\frac{\partial u_o}{\partial n}\, ds = \int_\gamma u_1 \frac{\partial u_o}{\partial n}\, ds$$

($u_o=0$ on γ). The last integral vanishes in view of (13) and the fact that, by the lemma and the definition of u_1, the latter function takes algebraically opposite values on opposite edges of γ. Accordingly, (14) reduces to

$$(u,u)_D = (u_o,u_o)_{D_o} + (u_1-u_o,\ u_1-u_o),$$

which shows that $(u,u)_D > (u_0,u_0)_{D_0}$, except when u and u_1 coincide.

This type of problem, may, of course, also be solved by means of variational techniques. However, in addition to its greater simplicity, the present procedure has the advantage of yielding sufficient conditions while it is in the nature of variational techniques to furnish necessary conditions only. This feature may be useful in settling questions of uniqueness. The weakness of the present method as compared with the Schiffer variation method lies in the fact that it does not yield the existence of the extremal domains. While in many cases -- as in that discussed above -- it is very easy to supply the required existence proof by means of simple considerations, there are problems where this is difficult and where the use of the present method will be restricted to showing the sufficiency of the necessary conditions furnished by variational procedures.

BIBLIOGRAPHY

[1] BERGMAN, S., The kernel function and conformal mapping, New York, American Mathematical Society, 1950.

[2] BERGMAN, S., and SCHIFFER, M., Kernel functions and conformal mapping, Compositio Math. 8 (1951), pp. 205-249.

[3] GRÖTZSCH, H., Über einige Extremalprobleme der konformen Abbildung, Berichte der sächs. Akad. d. Wiss. 80 (1928), pp. 367-376, pp. 497-502.

[4] GRUNSKY, H., Neue Abschaetzungen zur konformen Abbildung ein - und mehrfach zusammenhängender Bereiche, Schriften des math. Sem. d. Univ. Berlin, 1, Heft 3 (1932).

[5] GRUNSKY, H., Koeffizientenbedingungen für schlicht abbildende meromorphe Funktionen, Math. Zeits. 45 (1939), pp. 26-61.

[6] SCHIFFER, M., The kernel function of an orthonormal system, Duke Math. J., 13 (1946), pp. 529-540.

[7] SCHIFFER, M., Hadamard's formula and the variation of domain functions, Amer. J. Math. 68 (1946), pp. 417-448.

ON THE EFFECTIVE DETERMINATION
OF CONFORMAL MAPS

S. E. Warschawski

§1. INTRODUCTION

Suppose C is a closed Jordan curve with continuously turning tangent represented by the equation $z = z(s)$ where the arc length s is the parameter, $0 \leq s \leq L$. Suppose that $w = f(z)$ maps the interior R of C conformally onto the circle $|w| < 1$ so that $z = z_1$ in R and $z = z_2$ on C correspond to $w = 0$ and $w = 1$, respectively; let the definition of $f(z)$ be extended to C so that $f(z)$ is continuous in the closure of R. The function $\theta(s) = \arg f(z(s))$ satisfies the integral equation[1]

$$\theta(s) = \int_0^L K(s,t)\,\theta(t)dt - 2\beta(s) . \tag{1}$$

Here

$$\beta(s) = \arg \frac{z_2 - z(s)}{z_1 - z(s)}, \quad K(s,t) = \frac{\sin(\tau - \varphi)}{\pi r_{st}} = \frac{1}{\pi} \frac{\partial \log \frac{1}{r_{st}}}{\partial n_t}$$

where $\tau = \tau(t)$ is the tangent angle, $\varphi = \arg[z(t)-z(s)]$, $r_{st} = |z(t)-z(s)|$, and $\frac{\partial}{\partial n_t}$; indicates differentiation with resepct to the inner normal at $z(t)$.

We are interested here in the solution of (1) by iteration. The integral equation (1) is the same as the one for the Dirichlet problem in potential theory, and the solution of the integral equations for the Dirichlet and Neumann problems by means of suitable iterations is, of course, the classical approach of Carl Neumann and Poincaré to these problems. Inasmuch as these iteration processes are well suited for calculations on high speed computing machines it seems of interest to re-examine this method from this point of view.

Since $+1$ is the numerically least characteristic value of the kernel $K(s,t)$ and the existence of a solution of (1) is assured by Riemann's mapping theorem, the Neumann series converges and, in fact, is

[1] In this form the integral equation is due to S. Gershgorin [6] however, a similar integral equation was given earlier by L. Lichtenstein [9]. See also G. F. Carrier [4] and Garrett Birkhoff, D. M. Young, and E. H. Zarantello [2]. The integral exists as a principal value; if $\tau(s)$ satisfies a Holder condition then, of course, it exists as a Lebesgue integral.

dominated by

$$\sum_n \frac{1}{|\lambda_2|^n},$$

where λ_2 is the second characteristic value in order of magnitude. The novel part of this paper is an explicit estimate for the error after n iterations (i.e., the remainder of the Neumann series) and the derivation of lower bounds for $|\lambda_2|$, which are essential for the appraisal of the error. Finally, the degree of convergences is studied for special domains, such as "nearly convex" and "nearly circular" domains.[2]

This is a summary of the results; the complete paper, including some numerical examples, is to appear as a publication of the National Bureau of Standards.

§2. PROPERTIES OF THE KERNEL K(s,t).

We assume that the tangent angle $\tau(s)$ of C as function of the arc length s satisfies a Hölder condition with the exponent α, $0 < \alpha \leq 1$. It is well known that the kernel $K(s,t)$ possesses a discrete set of characteristic numbers $\lambda_1, \lambda_2, \ldots$

$$\lambda_1 = 1 < |\lambda_2| \leq |\lambda_3| \leq \ldots,$$

which are all real, and with every λ_i there is associated a pair of characteristic (continuous) functions $\psi_i(s)$ and $\varphi_i(s)$ which are solutions of the homogeneous integral equations

$$\psi_i(s) = \lambda_i \int K(s,t)\, \psi_i(t)\, dt$$

and

$$\varphi_i(s) = \lambda_i \int K(t,s)\, \varphi_i(t)\, dt,$$

respectively. The $\psi_i(s)$ and $\varphi_i(s)$ may be so chosen as to form a normalized bi-orthogonal set, i.e.

$$\int_C \varphi_i(s)\, \psi_j(s)\, ds = \delta_{ij} \tag{2}$$

where δ_{ij} is the Kronecker symbol. Furthermore, $\lambda_1 = 1$ is simple, $\psi_1(s) = \text{const.}$, and will be chosen $= 1$. Then

$$\int_C \varphi_1(s) ds = \int_C \varphi_1(s)\, \psi_1(s) ds = 1$$

[2] The present results are applicable with suitable modifications to the other integral equations of the Dirichlet and Neumann problems (for which $\lambda = 1$ is not a characteristic value).

and by (2)

$$\int_C \varphi_i(s)ds = 0 \quad \text{for} \quad i \geqq 2. \tag{3}$$

The kernel $K(s,t)$ is not symmetrical but it belongs to a class of symmetrizable kernels,[3] i.e., there exists a symmetrical function $\Phi(s,t) = \Phi(t,s)$ such that

$$H(s,t) = \int_C \Phi(s,x)\, K(t,x)dx$$

is symmetrical in s and t. Here $\Phi(s,t) = \log \frac{1}{r_{st}}$ where r_{st} denotes the distance between the points s and t of C. This fact permits one to characterize the eigenvalues λ_i and in particular λ_2 by a maximum property (as it is done in the case of a symmetrical kernel).

The integral

$$\int_C\int_C \log \frac{1}{r_{st}}\; \varphi(s)\; \varphi(t)dsdt > 0$$

for all continuous function $\varphi(t) \not\equiv 0$, for which

$$\int_C \varphi(t)dt = 0. \tag{4}$$

(It represents the energy of the charge $\varphi(s)ds$ distributed on C). <u>If</u>

$$H(s,t) = \int_C \log \frac{1}{r_{sx}}\, K(t,x)dx,$$

<u>then one has</u>

$$\frac{1}{|\lambda_2|} = \text{Max}\, \frac{\left|\int_C\int_C H(s,t)\; \varphi(s)\; \varphi(t)dsdt\right|}{\int_C\int_C \log \frac{1}{r_{st}}\, \varphi(s)\; \varphi(t)dsdt} \tag{5}$$

<u>for all continuous</u> $\varphi(t)$ <u>satisfying (4). The maximum is attained for</u> $\varphi_2(t)$.

This can be seen as follows: The space of all continuous functions $\varphi(t)$, $0 \leqq t \leqq L$, which satisfy (4) is completed to a Hilbert space whereby we use as the norm

$$||\varphi(t)||^2 = \int_C\int_C \log \frac{1}{r_{st}}\; \varphi(s)\; \varphi(t)dsdt, \quad (\varphi,\psi) = \int_C\int_C \log \frac{1}{r_{st}}\, \varphi(s)\; \psi(t)dsdt.$$

[3] See [3], [5], [7], [10], [11], [12] where integral equations with such a kernel were investigated.

Applying the standard proof for the existence of a characteristic value and function of a completely continuous Hermitian operator T in this Hilbert space, one shows that

$$\frac{1}{|\lambda_2|} = \sup_{||\varphi||\leq 1} |(T\varphi,\varphi)| .$$

This least upper bound is attained for a continuous function $\varphi_2(t)$. It is therefore sufficient in determining $\varphi_2(t)$ to restrict the range of functions φ to continuous functions and one arrives at (5).

More generally we shall write $||f(t)||^2$ for the integral

$$\int_C\int_C \log\frac{D}{r_{st}} f(s)f(t)dsdt$$

whenever the integral exists in the sense of Lebesgue.

The existence of a solution $\theta(s)$ of the integral equation (1) is assured by Riemann's mapping theorem. Since $\lambda = 1$ is a characteristic value of $K(s,t)$ it follows that

$$\int_C \beta(s)\ \varphi_1(s)ds = 0$$

Furthermore, the solution $\theta(s)$ is only determined up to an additive multiple of the characteristic function $\varphi_1(s) = 1$, i.e. up to an arbitrary additive constant.

§3. THE ITERATIONS. ESTIMATE OF THE ERROR

Let $\theta_0(s)$ possess a continuous derivative for $0 \leqq s \leqq L$. We form the sequence

$$\theta_{n+1}(s) = \int_0^L K(s,t)\theta_n(t)\ t - 2\,\beta(s), \qquad n = 0,1,2,\dots$$

and show that it converges uniformly in $0 \leqq s \leqq L$ to the solution $\theta(s)$ of the integral equation (1) which is characterized by the property

$$\int_0^L \theta(s)\varphi_1(s)ds = \int_0^L \theta_0(s)\varphi_1(s)ds. \tag{6}$$

This is contained in the following

THEOREM 1. If the tangent angle $\tau(s)$ of C, as function of the arc length s satisfies a Holder condition with the exponent α, $0 < \alpha \leqq 1$, and if $\theta_0(L) - \theta_0(0) = 2\pi$, then

$$|\theta_{n+1}(s) - \theta(s)| \leq$$

$$\frac{1}{\pi}\frac{1}{|\lambda_2|^{n+1}} \|K(s,t)\| \, \|\theta_0'(t) - \theta'(t)\| \left(\frac{\lambda_2^2}{\lambda_2^2 - 1}\right)^{1/2} \tag{7}$$

where $\theta(s)$ is the solution of (1) for which (6) holds

Here $|\lambda_2| > 1$; methods for obtaining an upper bound for $\frac{1}{|\lambda_2|}$ which is < 1 will be given below. $\theta'(s)$ exists and is continuous under the present assumption on $\tau(s)$. The factor $\|\theta_0'(t) - \theta'(t)\|$ in the estimate of the error is favorable, when the initial function $\theta_0(s)$ is a "good guess", i.e. when

$$\operatorname*{Max}_{0 \leq t \leq L} |\theta_0'(t) - \theta'(t)|$$

or even when only

$$\int_0^L \left(\theta_0'(t) - \theta'(t)\right)^2 dt$$

is a priori known to be "small". For, since by the inequality of Schwarz,

$$\|\theta_0'(t) - \theta'(t)\| \leq$$

$$\left\{\int_0^L \int_0^L \left(\log \frac{1}{r_{st}}\right)^2 ds dt\right\}^{1/2} \int_0^L \left(\theta_0'(t) - \theta'(t)\right)^2 dt$$

the factor $\|\theta_0'(t) - \theta'(t)\|$ will also be "small".

In the case where $\theta_0(s)$ is chosen arbitrarily we may use an a priori estimate for $\theta'(s)$; methods for finding such a bound are well known. The following estimate is derived in this paper by use of the integral equation (1). Under the assumption that the tangent angle $\tau(s)$ satisfies a Hölder condition of order α, it follows that

$$|K(s,t)| \leq \frac{A}{|s-t|^{1-\alpha}}$$

where A is a known constant. Let $\Phi(s) = \arg(z(s) - z_0)$ and

$$N = \operatorname*{Max}_{0 \leq s \leq L} |\Phi'(s)| .$$

Then

$$(8) \qquad 0 < \theta'(s) \leq \pi\left[\frac{LN}{\pi}\left(\frac{2}{L}\right)^{\alpha} + \frac{2A}{\alpha}\right]^{\frac{1}{\alpha}}, \qquad 0 \leq s \leq L.$$

It is, of course, not necessary to assume the knowledge of a solution $\theta(s)$, for it is also found by the iteration process, and one may eliminate $\theta(s)$ completely from the estimate of the error. The same method of proof used to establish (7) shows that

$$(9) \qquad |\theta_{n+1}(s) - \theta_n(s)| \leq \frac{1}{\pi}\frac{1}{|\lambda_2|^{n-1}} \, ||K(s,t)|| \, ||\theta_0'(t) - \theta_1'(t)|| \left(\frac{\lambda_2^2}{\lambda_2^2 - 1}\right)^{1/2} = \frac{B}{|\lambda_2|^{n-1}}$$

Consequently for

$$\theta(s) = \theta_0(s) + \sum_{k=0}^{\infty}\left(\theta_{k+1}(s) - \theta_k(s)\right)$$

one has

$$|\theta_{n+1}(s) - \theta(s)| \leq \frac{B}{|\lambda_2|^n}\,\frac{|\lambda_2|}{|\lambda_2| - 1}$$

The proof of (7) in Theorem 1 and (9) is based upon the use of certain expansion theorems of an arbitrary function into series of the form $\Sigma\, a_k\psi_k(s)$ and $\Sigma\, b_k\varphi_k(s)$ in terms of the characteristic functions $\psi_k(s)$ and $\varphi_k(s)$. These are due to Blumenfeld and Mayer [3] and to Carleman [5] for the case of the kernel $K(s,t)$.

§4. ESTIMATES OF $|\lambda_2|$

In order to calculate the errors in (7) and (9) it is still necessary to have an appraisal for $|\lambda_2|$. The estimates given here for $|\lambda_2|$ will be of the following type. Let Γ be a simple closed curve and $K_0(s,t)$ the kernel pertaining to Γ. Suppose that the second characteristic value μ_2 of $K_0(s,t)$ is known; for example for an ellipse with the semi-axes a and b, $a > b$, $\mu_2 = \frac{a+b}{a-b}$; for a circle $\mu_2 = \infty$. If Γ is (in a sense to be specified) close to C then we give an estimate of $\frac{1}{|\lambda_2|}$ in terms of $\frac{1}{|\mu_2|}$ In the following we write $\lambda = |\lambda_2|$, $\mu = |\mu_2|$

The characterization of $\frac{1}{\lambda}$ by the maximal property (5) may be put into a different form. Let $V(z)$ be the single layer potential

$$(10) \qquad V(z) = \int_C \varphi(t)\log\frac{1}{|z - z(t)|}dt \quad \text{with} \quad \int_C \varphi(t)dt = 0,$$

where $\varphi(t)$ is a continuous function. Let

$$D_1[V] = \iint_{I(C)} \left\{ \left(\frac{\partial V}{\partial x}\right)^2 + \left(\frac{\partial V}{\partial y}\right)^2 \right\} dxdy,$$

$$D_e[V] = \iint_{E(C)} \left\{ \left(\frac{\partial V}{\partial x}\right)^2 + \left(\frac{\partial V}{\partial y}\right)^2 \right\} dxdy$$

be the Dirichlet integrals for $V(z)$, taken over the interior and the exterior of C, respectively. Then (5) may be written in the form:

$$\frac{1}{\lambda} = \frac{1}{|\lambda_2|} = \text{Max} \frac{|D_e[V] - D_1[V]|}{D_e[V] + D_1[V]} \tag{11}$$

<u>for all potentials $V(z)$ of the form (10). The maximum is then attained when $\varphi(t) = \varphi_2(t)$.</u>

Using this characterization of λ we prove the following theorem.

THEOREM 2. Suppose that C and Γ are simple closed curves, C of class C_h'', Γ of class C_h',[4] and that either Γ is in the interior of C or C in the interior of Γ. Let

$$A = \int_0^L \left|\left|\frac{\partial K(s,t)}{\partial s}\right|\right|^2 dt,$$

$$m = \underset{0 \leqq s \leqq L}{\text{Max}} \; \theta'(s), \tag{12}$$

$$M = \frac{1}{\pi} \underset{0 \leqq s \leqq L}{\text{Max}} \frac{1}{\theta'(s)}$$

If d is the Frechet distance between C and Γ, then

$$\frac{1}{\lambda} \leqq \frac{1}{\mu} + d\lambda^2 a \quad (a = 2A\ m\ M)$$

If d is so small that $\frac{1}{\mu} + da < 1$, an upper bound for $\frac{1}{\lambda}$ will be given by the reciprocal of the root of the cubic equation

$$1 = \frac{x}{\mu} + dax^3$$

[4] We call a closed curve of class C_h' if the tangent angle and of class C_h'' if the curvature satisfies a Hölder condition.

which is greater than one.[5]

An upper bound for m is given in (8). Method for obtaining estimates for M are known in the literature, see for example [13], [16], and [14], [15] when C is "nearly circular".

Utilizing the maximum property of $\frac{1}{\lambda}$ in the form (5) we obtain another "comparison" theorem.

THEOREM 3. Suppose that C is of class C_h'', Γ of class C_h' and that both curves have the same length L. Suppose that (for some choice of the points corresponding to $s = o$ on each curve)

$$\int_0^L \int_0^L \left\{ K(s,t) - K_o(s,t) \right\}^2 dsdt = \varepsilon^2$$

$$\int_0^L \int_0^L \left\{ \log \frac{1}{r_{st}} - \log \frac{1}{\rho_{st}} \right\}^2 dsdt = \delta^2$$

where ρ_{st} denotes the distance of the points of Γ corresponding to the parameter values s and t and $K_o(s,t)$ is the kernel defined for Γ.

Then

$$\frac{1}{\lambda} \leqq \frac{1}{\mu} + \varepsilon A_1 \lambda^2 + \delta B_1 \lambda \mu \tag{13}$$

where A_1 and B_1 are constants

$$A_1 = \frac{2M}{\pi^2} \int_0^L \left\| \frac{\partial K(s,t)}{\partial s} \right\|^2 ds,$$

[5]In a paper [1] scheduled to appear soon L. Ahlfors deals with a generalization of the Neumann-Poincaré integral equation and obtains an estimate of a different nature for $|\lambda_2|$: He proves: if there exists a quasi conformal transformation of the interior of the curve C onto its exterior which leaves C fixed and whose maximal excentricity is $\leqq k<1$, then $\frac{1}{|\lambda_2|} \leqq k$. As an application he then shows that for a star-shaped curve C given in polar form $r = r(\varphi)$, $0 \leqq \varphi \leqq 2\pi$, $\frac{1}{|\lambda_2|} \leqq \text{Max } |\cos \gamma|$ where γ is the angle between the tangent and the radius vector at (φ, r). This second result furnishes a large class of comparison curves for our theorems 2 and 3. We should also like to mention the investigation [17] by S. Bergmann and M. Schiffer, in which the characteristic values λ_i appear as eigenvalues of a different integral equation in the complex domain, and the authors derive in this connection variational formulas for the λ_i (and the corresponding characteristic function) for infinitesimal deformations of the domain. However, they are not concerned with estimates for the change in the λ_i corresponding to a "finite" deformation.

$$B_1 = \left(\int_0^L \int_0^L (\log \frac{1}{\rho_{st}})^2 \, ds dt \right)^{1/2}$$

In (13) we may replace on the right hand side λ by μ, and use the resulting expression on the right as an upper bound for $\frac{1}{\lambda}$. A somewhat more favorable bound is again given by the reciprocal of the root of the cubic equation

$$1 = \frac{x}{\mu} + x^3 A_1(2\delta + B_1 \varepsilon).$$

which is greater than 1. (This root exists when $\frac{1}{\mu} + A_1(2\delta + B_1\varepsilon) < 1$).

Finally we mention an estimate of a different kind. It is easily established that for

$$K_2(s,t) = \int_C K(s,x)\, K(x,t) dx$$

$$\frac{1}{\lambda_2^2} \leq \frac{1}{2} \left\{ \int_0^L K_2(s,s) ds - 1 \right\}.$$

By the inequality of Schwarz we have then

$$\frac{1}{\lambda_2^2} \leq \left\{ \int_0^L \int_0^L K^2(s,t) dsdt - 1 \right\}$$

$$= \frac{1}{2} \int_0^L \int_0^L \left\{ K(s,t) - \frac{1}{L} \right\}^2 dsdt.$$

Thus, if the last integral is < 1, we obtain a usable estimate for $\frac{1}{|\lambda_2|}$. Since, for a circle of circumference L, the kernel $K_o(s,t) = \frac{1}{L}$ the condition that

$$\frac{1}{2} \int_0^L \int_0^L \left(K(s,t) - \frac{1}{L} \right)^2 dsdt < 1$$

may be interpreted as requiring that C be "nearly circular" in this sense.[6]

---

[6] In the case of "nearly circular" regions one might be interested in comparing the present method with that of Theodorsen and Garrick. The principal advantage of the integral equation (1) is the fact that the kernel $K(s,t)$ is simpler. For curves C with continuous curvature $K(s,t)$ is continuous, while the kernel of Theodorsen's integral equation is singular. This seems to simplify the numerical procedure.

§5. SPECIAL DOMAINS

For a special class of domains which will be denoted as "nearly convex" we estimate the degree of convergence of the iterations by a method which is a modification of a classical argument due to Carl Neumann for the case of convex curves. These results are of interest because of their simple form. If C is a convex closed curve then $K(s,t) \geqq 0$. We state first:

Neumann's Lemma. If C is a convex curve with continuously turning tangent then there exists a constant k, $0 < k < 1$ (the "Neumann constant") which depends only on C and has the following property: Let $g(t)$ be integrable along C, $0 \leqq t \leqq L$, and let

$$g^*(s) = \int_0^L g(t)\, K(s,t)dt.$$

If M and m are the least upper and greatest lower bounds of $g(t)$ in $[0,L]$ and if M^* and m^* denote the corresponding quantities for $g^*(s)$, then

$$M^* - m^* \leqq (M - m)\ (1-k).$$

The constant k may be estimated for certain convex curves.

We denote a rectifiable curve C as "nearly convex" if it satisfies the following condition: there exists a closed convex curve C_o with continuously turning tangent such that

(i) C_o has the same length L as C; and

(ii) If $K(s,t)$ and $K\ (s,t)$ are the kernels for C and C_o, respectively, then for all s, $0 \leqq s \leqq L$,

$$\int_0^L |K(s,t) - K_o(s,t)|dt \leqq \varepsilon < k. \tag{14}$$

THEOREM 4. If C is "nearly convex" and if $\theta_o(s)$ is an arbitrary continuous function, $0 \leqq s \leqq L$, then the iterations $\theta_n(s)$ defined in § 3 satisfy the inequalities:

$$|\theta_{n+1}(s) - \theta_n(s)| \leqq V(1+\varepsilon)\ (1-k+\varepsilon)^n, \quad 0 \leqq s \leqq L.$$

Here V is a constant,

$$V \leqq \Omega_o + 2\pi$$

and Ω_o is the oscillation of $\theta_o(s)$ in $[0,L]$.

If $\theta_0(s)$ is non-decreasing and $\Omega_0 = 2\pi$ then $V = 2\pi$. If $\theta_0(s)$ is an "approximation" to the solution i.e. if it is known a priori that, for some solution $\theta(s)$ of the integral equation (1),

$$\max_{0 \leq s \leq L} |\theta(s) - \theta_0(s)| \leq \eta$$

then

$$V \leq 2\eta.$$

For the case that C_0 is a circle, $k = 1$ we have another type of a "nearly circular" region, where (14) serves as a measure for the closeness of C to a circle.

BIBLIOGRAPHY

1. AHLFORS, L. V., Remarks on the Neumann-Poincaré Integral Equation, to appear in Pacific Journal of Mathematics.
2. BIRKHOFF, GARRET, YOUNG, D. M., ZARANTELLO, E. H., Effective conformal transformation of smooth simply connected domains, Proceedings, National Academy of Sciences, vol. 37, 1951, pp. 411-414.
3. BLUMENFELD, J., MAYER, W., Uber Poincarésche Fundamentalfunktionen, Sitzungsberichte, Wiener Akademie der Wissenschaften, Math.-Naturwissenschaftliche Klasse, Bd. 122, Abt.IIa, (1914), pp. 2011-2047.
4. CARRIER, G. F., On a conformal mapping technique, Quarterly Applied Mathematics, vol. 5 (1947), pp. 101-104.
5. CARLEMAN, T., Über das Neumann-Poincarésche Problem fuer ein Gebiet mit Ecken, Upsala Dissertation, 1917.
6. GERSHGORIN, S., On conformal mapping of a simply connected region onto a circle, (russian) Matematicheski Sbornik, vol. 40 (1933) pp. 48-58.
7. KORN, A., Über die erste und zweite Randwertaufgabe der Potentialtheorie, Rendiconti del Circolo Matematico di Palermo, vol. 25 (1913).
8. KORN, A., Über die Anwendung der Methode der Sukzessiven Näherungen zur Lösung von linearen Integralgleichungen mit unsymmetrischen Kernen, Archiv der Mathematik und Physik, Bd. 25, (1917) pp. 148-173 and Bd. 27.
9. LICHTENSTEIN, L., Zur Konformen Abbildung einfach zusammenhangender schlichter Gebiete, Archiv fur Mathematik und Physik, Bd. 25 (1917), pp. 179-180.
10. MARTY, J., Développements suivant certaines solutions singulières, Comptes Rendus, CL (1910) pp. 603-06.
11. MARTY, J., Existence de solutions singulières pour certaines équations de Fredholm, Comptes Rendus CL (1910), 1031.
12. PLEMELJ, J., Potentialtheoretische Untersuchungen, Teubner, Leipzig, 1911.
13. SEIDEL, W., Über Ränderzuordnung bei konformen Abbildungen, Mathem. Annalen Bd. 104, 1931, pp. 182-243.
14. SPECHT, E., Estimates of the mapping function and its derivatives in conformal mapping of nearly circular regions, Transaction, Am. Math. Society, vol. 71, (1951) pp. 183-196.
15. WARSCHAWSKI, S. E., On conformal mapping of nearly circular regions, Proceedings, Am. Math. Society, vol. 1 (1950) pp. 562-574.

16. WARSCHAWSKI, S. E., On conformal mapping of regions bounded by smooth curves, ibid., vol. 2, (1951), pp. 254-261.

17. BERGMAN, S., SCHIFFER, M., Kernel functions and conformal mapping, Compositio Mathematica, vol. 8 (1951), pp. 205-249.

STRUCTURE OF COMPLEX SPACES

S. Bochner

§1. COMPLEX AND REAL COORDINATE SPACES

The natural generalization of the classical Riemann surface from one to several variables is the connected complex coordinate space S_{2k} and we define it as follows, see [14], p. 54. It is a neighborhood space, locally Euclidean, of an even dimension $n = 2k$; and there are given a covering of it by a family of neighborhoods $\{U_\nu\}$, and topological mappings $V_\nu = \chi_\nu(U_\nu)$ into the ordinary space E_{2k} of the variables $z_\alpha = x_\alpha + iy_\alpha$, $\alpha = 1,\ldots,k$, such that if two neighborhoods U_μ, U_ν have a non-empty intersection U, and if we put $V = \chi_\mu(U)$, $V' = \chi_\nu(U)$, then the functions

$$x'_\alpha = \varphi_\alpha(x,y), \quad y'_\alpha = \psi_\alpha(x,y), \quad \alpha = 1,\ldots,k \tag{1}$$

which transmit the map $V' = \chi_\nu \chi_\mu^{-1}(V)$ shall be the real and imaginary parts of certain holomorphic functions

$$z'_\alpha = f_\alpha(z_1,\ldots,z_n) \quad \alpha = 1,\ldots,k. \tag{2}$$

We will call our space a <u>real</u> space if we do not assume that our functions (1) have this holomorphic structure, and we will then assume that they are twice differentiable say. Formally, by putting $\Phi_\alpha = \varphi_\alpha + i\,\psi_\alpha$, we can then always rewrite (1) into

$$z'_\alpha = \Phi_\alpha(z,\bar{z}), \quad \bar{z}'_\alpha = \bar{\Phi}_\alpha, \tag{3}$$

and our real space is also a complex space if and only if we have

$$\frac{\partial \Phi_\alpha}{\partial \bar{z}_\beta} = 0 = \frac{\partial \bar{\Phi}_\alpha}{\partial z_\beta}, \quad \alpha,\beta = 1,\ldots,k. \tag{4}$$

Furthermore, for the Jacobian of the transformation (1) we can always write

$$\frac{\partial(z'_1,\ldots,z'_k,\bar{z}'_1,\ldots,\bar{z}'_k)}{\partial(z_1,\ldots,z_k,\bar{z}_1,\ldots,\bar{z}_k)}, \tag{5}$$

see [14], p. 39, and since for a complex space (4) this is

$$\frac{\partial(z'_1,\ldots,z'_k)}{\partial(z_1,\ldots,z_k)} \cdot \frac{\partial(\bar{z}'_1,\ldots,\bar{z}'_k)}{\partial(\bar{z}_1,\ldots,\bar{z}_k)}, \tag{6}$$

it follows that a complex space is always orientable, perforce.

§2. SEPARABILITY

For $k = 1$, T. Rado [17] has shown that our (connected) complex space has a covering by countably many complex neighborhoods automatically. It appears from many indications that for $k \geq 2$ complex structure virtually ceases to have influence on "separability" and we will briefly reproduce Rado's reasoning, in somewhat altered form, in order to show when and why it ceases to be applicable.

THEOREM 1. Let S be any arcwise connected neighborhood space and let $\{S_\alpha\}$ be the family of its perfectly separable arcwise connected open subspace containing a given point. Let E be another such space, and let it be possible to associate with every S_α a perfectly separable open subspace D_α of E such that for $S_\alpha > S_\beta$, $S_\alpha \neq S_\beta$ we have $D_\alpha > D_\beta$, $D_\alpha \neq D_\beta$.
Now, if E itself is perfectly separable, the S itself is likewise perfectly separable.

In fact, if we form the point set

$$D_\infty = \Sigma_\alpha D_\alpha \quad ,$$

then due to Lindelof there is a countable subset $\{D_n\}$, $n = 1,2,\ldots,$ such that

$$D_\infty = \sum_{n=1}^{\infty} D_n,$$

and if we take their preimages $\{S_n\}$ and introduce their union

$$S_\infty = \sum_{n=1}^{\infty} S_n$$

then the latter must be maximal among the separable subspaces of S introduced, but a maximal such element must be S itself, q.e.d.

Now, in Rado's case, S is a connected complex space, and E is the Gaussian sphere, and the D_α's are concentric circles, finite or infinite, or E itself. If now with each S_α we associate that D_α which arises by conformal mapping of its universal covering space by holding the common point and a given line-element of it fixed, then Theorem 1 applies and Rado's assertion follows. However, for $k \geq 2$, Theorem 1 ceases to be applicable towards the same end, not because of any delicate features of the uniformization theory failing to hold, but already on the crude grounds that purely topologically, even among algebraic surfaces only, there are too many types to fit as "subspaces" D_α into a single one, as Theorem 1 would require them to do.

§3. WHEN A REAL SPACE IS ALSO A COMPLEX SPACE

For $k = 1$, an orientable real space is also automatically a complex space if and only if it carries a conformal structure, and we are going to generalize this to $k \geq 2$, first apparent obstacles notwithstanding.

For $k = 1$, conformal structure is a line element which in every coordinate system V_α has the form

$$ds^2 = \lambda(x,y)\ (dx^2 + dy^2) = g(z,\overline{z})dz\ d\overline{z}, \tag{7}$$

and for $k \geq 2$, even if S_{2k} is compact, it is probably not always possible to have a line element of the form

$$\lambda(x,y) \sum_{\alpha=1}^{k} (dx_\alpha^2 + dy_\alpha^2) = g(z,\overline{z}) \sum_{\alpha=1}^{k} dz_\alpha\ d\overline{z}_\alpha\ . \tag{8}$$

However, as we have recently pointed out in [9], if S_{2k} has a countable covering by complex neighborhoods, we can always introduce a line-element having the Hermitian form

$$ds^2 = \sum_{\alpha,\beta=1}^{k} g_{\alpha\overline{\beta}}\ dz_\alpha d\overline{z}_\beta\ , \qquad g_{\beta\overline{\alpha}} = \overline{g_{\alpha\overline{\beta}}} \tag{9}$$

and this is our generlization of (7).

Now, conversely, for $k = 1$, if a real space is orientable, and carries a metric (7) then it is automatically complex. For $k \geq 2$, even if a real space is orientable and compact, then the presence of a conformal metric (8), and let alone of an Hermitian metric (9), will not make it complex, as already disproved by the ordinary four-sphere. However the addition to (9) of a certain structural requirement which for a complex space is fulfilled anyway, will so make it, always.

THEOREM 2. If a real space is such that for the functions (3) we have

$$\det \left| \frac{\partial \Phi_\alpha}{\partial z_\beta} \right|_{\alpha,\beta = 1,\ldots,k} \neq 0 \tag{10}$$

at all points, and if a non-singular Hermitian metric (9) is known to exist (it need not be definite) then the space is automatically a complex one.

In fact, the equality

$$\Sigma\, g_{\alpha\bar{\beta}}\, dz_{\alpha}\, d\bar{z}_{\beta} = \Sigma\, g'_{\gamma\bar{\delta}}\, dz'_{\gamma}\, d\bar{z}'_{\delta} \tag{11}$$

will imply the identities

$$\begin{aligned} g_{\alpha\bar{\beta}} &= g'_{\gamma\bar{\delta}} \frac{\partial \Phi_{\gamma}}{\partial z_{\alpha}} \frac{\partial \bar{\Phi}_{\delta}}{\partial \bar{z}_{\beta}} \\ 0 &= g'_{\gamma\bar{\delta}} \frac{\partial \bar{\Phi}_{\gamma}}{\partial z_{\alpha}} \frac{\partial \Phi_{\delta}}{\partial \bar{z}_{\beta}} \end{aligned} \tag{12}$$

and their conjugate complex ones, and for a non-singular matrix $|g'_{\gamma\bar{\delta}}|$ together with (10), relations (12) will imply relations (4), as claimed.

Next, on a real space, an object with coordinates $(\zeta_1, \ldots, \zeta_k)$ will be called a (covariant) vectorfield if we have

$$\sum_{\alpha=1}^{k} \zeta_{\alpha}\, dz_{\alpha} = \sum_{\beta=1}^{k} \zeta'_{\beta}\, dz'_{\beta} \quad , \tag{13}$$

but unless postulated explicitly the components need not be holomorphic in the given coordinates, in any manner.

THEOREM 3. If on a real space there are given r vectorfields

$$\left\{ \zeta^{(\rho)}_{\alpha} \right\}, \qquad \alpha = 1, \ldots, k; \qquad \rho = 1, \ldots, r, \tag{14}$$

$r \geq k$, and if the rank of the matrix (14) has always its maximal value k, then the space is also complex.

In fact, (13) implies

$$0 = \sum_{\beta=1}^{k} \zeta'_{\beta} \frac{\partial \Phi_{\beta}}{\partial \bar{z}_{\alpha}}, \qquad \alpha = 1, \ldots, k,$$

and if this holds for r vectorfields of rank k then we obtain (4) as claimed.

Note, that our vectorfields generate the Hermitian metric tensor

$$g_{\alpha\bar{\beta}} = \sum_{\rho=1}^{r} \zeta^{(\rho)}_{\alpha}\, \overline{\zeta^{(\rho)}_{\beta}} \tag{15}$$

and we might have attempted to obtain the conclusion of Theorem 3 by way of Theorem 2 already proven. But this would have necessitated adding assumption (10), which in the direct proof of Theorem 3 is not required, however.

§4. RIGIDITY

We will draw a geometric conclusion, for $k \geq 2$, from Hartog's Theorem that if in $E_{2k} : (z_1, \ldots, z_k)$ a function is holomorphic in a connected neighborhood of a connected boundary of a bounded domain, then it can be continued (uniquely) into the entire domain; and we also note that in [1] and [13] we have generalized this to analytic functions in real variables which satisfy an equation

$$\Sigma_{0 \leq p_1 + \ldots + p_n \leq 2h} \; a_{p_1 \ldots p_n} \frac{\partial^{p_1 + \ldots + p_n} f}{\partial x_1^{p_1} \ldots \partial x_n^{p_n}} = 0$$

in all variables, and an additional equation

$$\Sigma_{0 \leq q_1 + \ldots + q_m \leq m} \; b_{q_1 \ldots q_m} \frac{\partial^{q_1 + \ldots + q_m} f}{\partial x_1^{q_1} \ldots \partial x_m^{q_m}} = 0$$

in fewer than all variables,

$$m < n,$$

both equations with constant coefficients, the second otherwise unrestricted, and the first being elliptic semi-homogeneous as defined in [13].

An immediate conclusion from Hartog's own theorem is that if we can map in E_{2k} the neighborhood of one boundary holomorphically into the neighborhood of another boundary then there exists a unique continuation of the map which carries the domain bounded into the domain bounded, and the reader will see that this can be interpreted in the following manner.

THEOREM 4. For $k \geq 2$, if in a complex S_{2k} we remove a "cap" lying all in one complex neighborhood and replace it by another such cap holomorphically, then the new cap is identical with the old one, unless, of course, it is possible to reorganize the complex structure altogether so as to identify points near the surface of excision with points away from it.

However, it is know already from algebraic geometry that it is very well possible to replace the simple cap by a mitre-like cover which arises from the simple cap by a birational transformation in which all points are mapped one-one except for the center of the cap which is being inflated into a hypersurface P_{2k-2} of the type of projective space. But the mitre is no longer a topological cell, and the problem remains of deciding whether it is possible to replace the cap by a holomorphically inequivalent topological cell also.

§5. KAEHLER METRIC GENERATED BY TENSOR FIELDS

A Kaehler metric is an Hermitian metric (9) with the added condition

$$\frac{\partial g_{\alpha\bar{\beta}}}{\partial z_\gamma} = \frac{\partial g_{\gamma\bar{\beta}}}{\partial z_\alpha} \tag{16}$$

and its conjugate complex; for a detailed discussion, compare [3] and [2], [6]. For $k = 1$, (16) is automatically fulfilled, for $k \geq 2$ there are compact complex manifolds on which no Kaehler metric can exist.

A Kaehler metric certainly does exist if we are given r vector-field as in Theorem 3, but this time holomorphic, and with the curl-property

$$\frac{\partial \zeta_\alpha^{(\rho)}}{\partial z_\beta} = \frac{\partial \zeta_\beta^{(\rho)}}{\partial z_\alpha} \tag{17}$$

added. In fact, the metric (15) already introduced will now have the Kaehler property in consequence of (17); and the meaning of (17) is that we can introduce the r simple Abelian integrals of the first kind

$$w_\rho(z) = \int^z \sum_{\alpha=1}^{k} \zeta_\alpha^{(\rho)}\, dz_\alpha , \tag{18}$$

which although perhaps not uniquely determined globally (they may have additive periods) are nevertheless creating locally a one-one holomorphic immersion of our S_{2k} into the E_{2r} : $(w_1, \ldots, w_r)$. The metric (15) is then induced by the metric

$$dw_1\, d\bar{w}_1 + \ldots + dw_r\, d\bar{w}_r \tag{19}$$

in E_{2r}, and since the latter is Kaehlerian, (15) is likewise so. Also, the metric (15) has then non-positive Ricci-curvature

$$-R_{\alpha\bar{\beta}} \leq 0, \tag{20}$$

see [3], p.190; and if S_{2k} is also compact, then (18) maps it locally one-one into a certain compact complex multi-torus, and the Euler characteristic of S_{2k} has then algebraic sign $(-1)^k$ or 0,. see [7], p. 255/6.

A parallel but different situation arises if the vector fields given are not covariant but contravariant, thus

$$\left\{ \zeta^\alpha_{(\rho)} \right\} \tag{21}$$

again with maximum rank k everywhere. The matrix

$$g^{\alpha\bar{\beta}} = \sum_{\rho=1}^{r} \zeta^{\alpha}_{(\rho)} \overline{\zeta^{\beta}_{(\rho)}} \tag{22}$$

is again positive definite Hermitian, but the requirement that its inverse $g_{\alpha\bar{\beta}}$ shall have the Kaehler property (16) is this time a very restrictive one, see [10] and [2]. It more or less demands that we have

$$\zeta^{\beta}_{(\rho)} \frac{\partial \zeta^{\alpha}_{(\sigma)}}{\partial z_{\beta}} - \zeta^{\beta}_{(\sigma)} \frac{\partial \zeta^{\alpha}_{(\rho)}}{\partial z_{\beta}} = 0, \tag{23}$$

$\alpha = 1, \ldots, k$; $\rho, \sigma = 1, \ldots, r$, but it turns out that the metric is then a flat metric, with the space being spread out over the flat E_{2k}, and, if compact, being a complex multitorus directly, see [10].

If we wish to construct a Kaehler metric by immersing S_{2k} holomorphically into a projective P_{2s-2} with homogeneous coordinates

$$w_1 : w_2 \cdots : w_s, \tag{24}$$

$s \geq k + 1$, it is no longer feasible to state conditions, for this to be possible, in terms of the vectorial gradients of the coordinates w_{σ} only, but we will also have to involve these coordinates functions themselves, undifferentiated. Since only the ratios of the coordinates count, the coordinates as objects on S_{2k} need not be holomorphic scalar functions proper, but it suffices that they be holomorphic scalar densities of some joint weight g, no matter what the weight. Such a density is an object having in each coordinate system a holomorphic component $\varphi(z)$ with a law of transformation

$$\varphi'(z') = \left(\frac{\partial(z)}{\partial(z')}\right)^{g} \varphi(z); \tag{25}$$

and if we are given s such densities $\varphi_{\sigma}(z)$ for which the matrix

$$\left| \varphi_{\sigma}, \frac{\partial \varphi_{\sigma}}{\partial z_{\alpha}} \right| \tag{26}$$

of s rows and $k+1$ columns has everywhere its maximal rank $k+1$, then we obtain an immersion into P_{2s-2} and thus a Kaehler metric on S_{2k}, compare [10].

And as regards securing sufficiently many densities of the same weight, the weight unlimited, we have noted in [10] that they may be obtained if there are given sufficiently many tensorfields

$$\zeta^{\beta_1 \cdots \beta_q}_{\alpha_1 \cdots \alpha_p} \tag{27}$$

or even tensor densities, of arbitrary type (p,q,g) by forming certain determinants with their components. Thus we obtain the following indeterminate proposition which could be made precise by drawing up specific assumptions laboriously.

THEOREM 5. A complex S_{2k} can be immersed holomorphically one-one into a projective space, and thus carries a Kaehler metric, if there are on it sufficiently many suitably independent holomorphic tensor densities (27) of any given type.

§6. UNIFORMIZATION

We will say that a complex space S_{2k} can be uniformized, if its universal covering space is holomorphically equivalent with a domain D in complex E_{2k}.

THEOREM 6. (Heuristic) If S_{2k} can be uniformized then it has a Kaehler metric for which the Ricci curvature is negative.

THEOREM 7. (conjectural, dubious) If S_{2k} has a Kaehler metric for which the Ricci curvature is negative, it can be uniformized.

THEOREM 8. (conjectural) If a complex space S_{2k} can be uniformized, then a lower dimensional complex space $S_{2\ell}$, $\ell < k$ which can be imbedded in it holomorphically locally one-one can be likewise uniformized.

As in [3] and [4], for the motivation of Theorem 6 we take as our prospective covering space a bounded domain D in E_{2k} on which there is a transitive group G of complex homeomorphisms

$$z'_j = f_j(z, \theta), \tag{28}$$

$\theta \in G$. With a Haar measure we introduce, if the integrals converge absolutely, the metric form

$$g_{\alpha\bar{\beta}} = \int_G \left(\sum_{j=1}^{k} \frac{\partial f_j(z, \theta)}{\partial z_\alpha} \cdot \overline{\frac{\partial f_j(z, \theta)}{\partial z_\beta}} \right) d\theta, \tag{29}$$

and we note that this metric is group-invariant, Kaehlerian, and semi-definite. If we add the explicit assumption that it is strictly positive definite it follows rigorously that its Ricci curvature is non-positive,

$$-R_{\alpha\bar{\beta}} \leq 0, \tag{30}$$

this following from the fact that the metric (29) can be interpreted as a limit of metrics each of which arises from an imbedding in a Euclidean space with a metric (19). Actually, in most interesting cases the Ricci curvature is strictly negative, and furthermore the metric is an Einstein metric

$$-R_{\alpha\bar{\beta}} = bg_{\alpha\bar{\beta}} \quad , \tag{31}$$

and finally the four-indices curvature itself is likewise negative.

Now, take a discrete subgroup Γ of G and form the space $S_{2k} = D/\Gamma$ of which D is the covering space. Since (29) is invariant with respect to G it gives rise to a Kaehler metric on S_{2k} itself, and (30) holds, as claimed.

As regards Theorem 7 we note that by a general theorem of (Hadamard and) E. Cartan a real space with Riemann metric and non-positive four-indices curvature has a covering space which differentiably is a Euclidean domain, and we are setting down the conjecture that for a complex space with Kaehler metric the weaker assumption (30) will produce the even stronger conclusion that the covering space is even holomorphically a domain.

Theorem 8 is now a combination of 6 and 7. In fact if S_{2k} is uniformizable then (30) holds; for an imbedded $S_{2\ell}$ we have then (30) even more so [this is a rigorous theorem, see [3], p. 190] and thus, conversely, $S_{2\ell}$ is likewise uniformizable.

We further note that a compact S_{2k} whose universal covering space is a <u>bounded</u> domain D has itself as a rule only a finite number of complex homeomorphisms, see [4] and also [16]. Thus, if it is an algebraic variety, as it probably always is, the field of meromorphic functions on it has only a finite number of automorphisms.

Also, see [2], (30) implies that there exists on it no holomorphic tensor field with only upper indices

$$\zeta^{\alpha_1 \ldots \alpha_p}$$

and if in addition it is an Einstein space (31) then there even exists no mixed holomorphic tensor field

$$\zeta^{\alpha_1 \ldots \alpha_p}_{\beta_1 \ldots \beta_q}$$

in which the number of upper indices predominates, $p > q$, see [11].

§7. COMPLEX MANIFOLDS WITH ALGEBRAIC SINGULARITIES

We are now going to give a brief extract from a forthcoming paper [15] jointly with W. T. Martin.

Consider, say over

$$|z_1|^2 + \dots + |z_k|^2 < 1, \tag{32}$$

a system of polynomial equations

$$P_m(\zeta_m, \zeta_{m-1}, \dots, \zeta_1; \quad z_1, \dots, z_k) = 0, \; m = 1,\dots,n, \tag{33}$$

in variables $\zeta_1, \dots, \zeta_n$ with coefficients holomorphic in (32) such that in the m-th polynomial the highest occurring power of ζ_m has total coefficient 1. We also assume that the system (33) is "irreducible" in the sense that there exists in (32) a certain holomorphic function

$$\Delta(z_j) = \Delta_U(z_j),$$

which function we choose and hold fast, such that for z not satisfying

$$\Delta(z_j) = 0 \tag{34}$$

the systems of solutions $(\zeta_1, \dots, \zeta_n)$ of (33) are all different from one another, and constitute themselves into holomorphic functional elements

$$\zeta_1 = \zeta_1(z), \dots, \zeta_n = \zeta_n(z) \tag{35}$$

which are freely continuable along any path for which $\Delta(z_j) \neq 0$ throughout.

If now we consider the system (33) as a system of equations in the space E_{2k+2n} of the complex variables (z_j, ζ_m), for z in (32) and all such ζ's as will solve the equations, then we obtain a well defined point set in E_{2k+2n}, and we may impart to it the topology of the latter space. We do this, and if so topologized, we denote our pointset by U; and we emphasize that this topologization embraces all points of our pointset without any exception. After this, we select in U those points whose z-coordinates satisfy (34) and we denote this subset of U by $N(U, \Delta)$. Now, the difference set

$$U - N(U, \Delta)$$

is obviously such that the functions (35) may be employed towards endowing it with a complex structure as defined in Section 1, and we so endow it.

If now we wish to define a "complex manifold S_{2k} with singularities" globally, we can do this in the following way. It is a topological space together with a covering by open sets U_α each of which is first of all topologically of the kind introduced. With each U_α there is associated a function Δ_α as before, and we are endowing the subspace

$$U_\alpha - N(U_\alpha, \Delta_\alpha) \tag{36}$$

with the complex structure as before. And, what is decisive, we assume that the union of sets

$$\Sigma_\alpha \left\{ U_\alpha - N(U_\alpha, \Delta_\alpha) \right\} , \tag{37}$$

which is a subspace of S_{2k}, is locally Euclidean, and is endowable with a complex structure globally in such a manner that the individual complex structures of the parts (36) shall be holomorphically compatible with it.

We say that a function $f(z)$ is holomorphic on such a S_{2k} if it is defined and continuous everywhere and if on every U_α it has a representation by an expression

$$\Sigma_{(\nu)} \zeta_1^{\nu_1} \cdots \zeta_n^{\nu_n} f_{\nu_1 \cdots \nu_n}(z), \tag{38}$$

$\nu_m \geq 0$, which is polynomial in the ζ's with coefficients which are holomorphic in (32).

A first theorem is that if a function $f(z)$ is defined in (37) only, and is bounded in every (36), and if there is a representation (38) in every (36), then it has a holomorphic continuation into all of S_{2k}. Furthermore, a holomorphic function in a subdomain assumes the maximum of its modulus on the boundary, so that in particular on a compact such S_{2k} it must be a constant. Next, there are satisfactory generalizations of Schwarz' lemma, and of other convexity theorem. And finally, and this is the least obvious generalization, it is again true that the uniform limit of holomorphic functions is again a holomorphic function.

If we wish likewise to extend the notions of holomorphic vector and tensor fields to our new manifolds, the exceptional points

$$\Sigma_\alpha N_\alpha (U_\alpha, \Delta_\alpha)$$

included, then the experience from algebraic geometry shows that in the (z, ζ) variables certain singularities for the components must be envisaged, if the objects are not to be restricted unduly. This can be done adequately and leading theorem may be then generalized. We can for instance retain the theorem, recently proven in [9] that on a compact S_{2k} the linear space of holomorphic tensor fields (27) of any one type (p,q,g) has again a finite basis, always.

§8. AUTOMORPHIC FUNCTIONS

Perhaps the most notable generalization possible, from regular manifolds to manifolds with singularities, is for a basic theorem affecting automorphic functions so-called, and the following very comprehensive version of it, that was recently established by ourselves in [12] goes over directly. If S_{2k} is compact and $\tilde{S}_{2k}$ is its universal covering space, then we call $f(P)$ automorphic if it is defined and holomorphic on $\tilde{S}_{2k}$ and for any element γ of the fundamental group we have

$$f(\gamma P) = \eta_\gamma(P)\, f(P),$$

where each factor of automorphy $\eta_\gamma(P)$ is itself holomorphic on $\tilde{S}_{2k}$. Now, take such a system of factors $\eta_\gamma(P)$ and consider $k + 2$ functions

$$f_o, f_1, \ldots, f_{k+1}$$

each of which is automorphic relative to these given factors, and the theorem is that there exists a non-vanishing polynomial with constant coefficients $Q(w_o, w_1, \ldots, w_{k+1})$ such that

$$Q\Big(f_o(P),\ f_1(P),\ \ldots,\ f_{k+1}(P)\Big) = 0.$$

This is the theorem, and the main application is that for $f_o(P) \neq 0$, the $k+1$ meromorphic quotients

$$g_\ell(P) = \frac{f_1\ (P)}{f_o\ (P)}$$

are connected by an (inhomogeneous) polynomial equation, so that, if k among the latter functions are not so connected, our space S_{2k}, singularities and all, may be viewed as the locus of an algebraic surface as defined in algebraic geometry, in a certain sense.

BIBLIOGRAPHY

[1] Analytic and meromorphic continuation by means of Green's formula, Annals of Math., 44 (1943), 652-673.

[2] Vector fields and Ricci curvature, Bull. American Math. Soc., 52 (1946), 776-797.

[3] Curvature in Hermitian metric, Bull. American Math. Soc., 53 (1947), 179-195.

[4] On compact complex manifolds, Jour. Indian Math. Soc., 51 (1947), 1-21.

[5] Curvature and Betti numbers, Annals of Math., 49 (1948), 379-390.

[6] Curvature and Betti numbers II, Annals of Math., 50 (1949), 77-93.

[7] Euler-Poincaré characteristic for locally homogeneous and complex spaces, Annals of Math., 51 (1950), 741-761.

[8] Vector fields on complex and real manifolds, Annals of Math., 52 (1950), 647-649.

[9] Tensor fields with finite basis, Annals of Math., 53 (1951), 400-411.

[10] Complex spaces with transitive commutative groups of transformations, Proc. Nat. Acad. Sci., 37 (1951), 356-359.

[11] Tensor fields and Ricci curvature in Hermitian metric, Proc. Nat. Acad. Sci., 37 (1951), 704-706.

[12] Algebraic and linear dependence of automorphic functions in several variables, Jour. Indian Math. Soc., 16 (1952), 1-6.

[13] Partial differential equations and analytic continuation, Proc. Nat. Acad. Sci., 38 (1952), 227-30

[14] BOCHNER, S., MARTIN, W. T., Several Complex variables, Princeton University Press, 1948. 216 pp.

[15] BOCHNER, S., MARTIN, W. T., Complex spaces with singularities, (to appear).

[16] HAWLEY, N. S., A theorem on compact complex manifolds, Annals of Math., 52 (1950), 637-641.

[17] RADO, T., Über den Begriff der Riemannschen Fläche, Acta Szeged, 2 (1925), 101-121.

REAL AND COMPLEX OPERATORS ON MANIFOLDS

D. C. Spencer*

On a Riemannian manifold of dimension m, the operator d of exterior differentiation maps forms of degree p into forms of degree $p + 1$, and the adjoint operator $\delta = (-1)^{mp+m+1} * d *$ maps forms of degree p into forms of degree $p - 1$. On a complex Kähler manifold each of these operators applied to forms of type (ρ,σ) can be split into the sum of two conjugate operators: $d = \partial + \bar\partial$, $\quad = \vartheta + \bar\vartheta$. The operator ∂ maps p-forms of type (ρ,σ), $\rho + \sigma = p$, into (p+1)-forms of type $(\rho + 1,\sigma)$ while its conjugate $\bar\partial$ maps p-forms of type (ρ,σ) into (p+1)-forms of type $(\rho,\sigma + 1)$. Dually, ϑ sends p-forms (ρ,σ) into (p-1)-forms $(\rho,\sigma - 1)$ while $\bar\vartheta$ sends p-forms (ρ,σ) into (p-1)-forms $(\rho - 1,\sigma)$.

A p-form φ is said to define a harmonic field if $d\varphi = \delta\varphi = 0$. On a Kähler manifold there are in addition complex fields defined by p-forms φ of type (ρ,σ) which satisfy $\partial\varphi = \bar\vartheta\varphi = 0$ (or $\bar\partial\varphi = \vartheta\varphi = 0$). If the Kähler manifold is compact (closed), a p-form φ of type (ρ,σ) which satisfies $\partial\varphi = \bar\vartheta\varphi = 0$ also satisfies $d\varphi = \delta\varphi = 0$ and conversely. But if the manifold is non-compact, the distinction of the two fields is significant; for example, a scalar satisfying $d\varphi = 0$ is a constant while one satisfying $\bar\partial\varphi = 0$ is a holomorphic function.

It is natural to try to extend the boundary-value problems of classical potential theory to subdomains of Riemannian manifolds, and it is also natural to consider the analogous problems for the complex operators on Kähler subdomains. For example, the complex analogue of Neumann's problem is that of finding a p-form φ of type (ρ,σ) which satisfies a differential equation in the interior of the domain and satisfies on the boundary the condition that the normal components of $\bar\partial\varphi$ (or $\partial\varphi$) assume prescribed values. In this connection we remark that a harmonic form φ of type (ρ,σ) with normal components of $\bar\partial\varphi$ vanishing on the boundary is holomorphic.

Real and complex boundary-value problems on manifolds were recently investigated ([7a,b]). Here we establish connections between these and other boundary-value theorems and various kernels occurring in formulas of

* This work was carried out under Office of Ordnance Research, U. S. Army Contract No. DA-36-034-ORD-639 RD.

orthogonal decomposition. The relationships obtained are generalizations of corresponding ones in the theory of functions, and they illustrate the duality of the real and complex operators.

The importance of introducing complex operators in a systematic way is that they generalize to higher-dimensional manifolds with boundary the essential formalism of classical complex potential theory. Although potential theory and complex structure on manifolds with boundary have a more tenuous connection than they have on closed manifolds, complex potential theory may still prove to be a useful tool in the investigation of complex structure on open manifolds - an investigation which is still in its infancy. Remarkable progress has been made in recent years in the understanding of complex structure on closed manifolds (Hodge, de Rham, H. Hopf, Bochner, Kodaira, Weil and others), and on subdomains of complex euclidean k-space (Bergman, Behnke, Stein, Oka and others). But at the present time almost nothing is known concerning complex function-theory on abstract manifolds of higher dimension which are not closed. An investigation of complex boundary-value problems on finite Kähler manifolds is an obvious initial step in the study of meromorphic tensors on non-closed manifolds.

1. COMPLEX OPERATORS

Let M be a complex manifold of complex dimension k with local analytic coordinates $z^1, z^2, \ldots, z^k$. We write

$$z^{i+k} = \bar{z}^i, \quad i = 1,2,\ldots,k,$$

and introduce the self-conjugate coordinates $z^1, z^2, \ldots, z^{2k}$. On M there is a real tensor $h_i{}^j$ satisfying

$$h_i{}^r \; h_r{}^j = \begin{cases} -1, & i = j, \\ 0, & i \neq j, \end{cases}$$

which, in self-conjugate coordinates, has the components

$$h_i{}^j = \begin{cases} \sqrt{-1}, & 1 \leq i = j \leq k, \\ -\sqrt{-1}, & k+1 \leq i = j \leq 2k, \\ 0, & i \neq j. \end{cases}$$

Let D_r denote the operator of covariant differentiation. The manifold is a Kähler manifold if it carries a positive-definite metric g_{ij} satisfying

$$g_{ij} = g_{rq} h_i^{\,r} h_j^{\,q}, \qquad D_r(h_i^{\,j} \varphi_j) = h_i^{\,j} D_r \varphi_j .$$

We write

$$h_{ij} = g_{jr} h_i^{\,r}, \qquad \omega = h_{(ij)}\, dz^i \wedge dz^j .$$

Here and elsewhere parentheses surrounding indices $i_1,\ldots,i_p$ indicate that the summation is restricted by the condition that $i_1 < i_2 < \ldots < i_p$. The Kähler restrictions imply that g_{ij} is a Hermitian metric satisfying $d\omega = \delta\omega = 0$.

We write

$$g_{i_1\ldots i_p,j_1\ldots j_p} = \begin{vmatrix} g_{i_1 j_1} \ldots g_{i_p j_1} \\ \\ g_{i_1 j_p} \ldots g_{i_p j_p} \end{vmatrix}$$

and, as in reference [3], we define

$$\underset{1,0}{\Pi}{}_i^{\,j} = \tfrac{1}{2}\,(g_i^{\,j} - \sqrt{-1}\, h_i^{\,j}),$$

$$\underset{0,1}{\Pi}{}_i^{\,j} = \underset{1,0}{\overline{\Pi}}{}_i^{\,j} = \tfrac{1}{2}\,(g_i^{\,j} + \sqrt{-1}\, h_i^{\,j}),$$

$$\underset{\rho,\sigma}{\Pi}{}_{i_1\ldots i_p}^{\;\;j_1\ldots j_p} = g_{i_1\ldots i_p}^{\;\;m_1\ldots m_\rho\, n_1\ldots n_\sigma}\, \underset{1,0}{\Pi}{}_{m_1}^{\,r_1} \cdots \underset{1,0}{\Pi}{}_{m_\rho}^{\,r_\rho}$$

$$\underset{0,1}{\Pi}{}_{n_1}^{\,s_1} \cdots \underset{0,1}{\Pi}{}_{n_\sigma}^{\,s_\sigma}\, g_{(r_1\ldots r_\rho)(s_1\ldots s_\sigma)}^{\;\;j_1\ldots j_p}$$

where ρ, σ are non-negative integers whose sum is p. Any p-form

$$\varphi = \varphi_{(i_1\ldots i_p)}\, dz^{i_1} \wedge dz^{i_2} \wedge \ldots \wedge dz^{i_p}$$

plainly has the orthogonal decomposition

$$\varphi = \sum_{\rho+\sigma=p} \prod_{\rho,\sigma} \varphi$$

where $\prod_{\rho,\sigma} \varphi$ is the p-form whose coefficients are

$$\prod_{\rho,\sigma}{}_{i_1 \ldots i_p}^{(j_1 \ldots j_p)} \varphi_{(j_1 \ldots j_p)} .$$

If $\varphi = \prod_{\rho,\sigma} \varphi$, we say that φ is of type (ρ,σ).

Now set

$$\partial = \prod_{\rho+1,\sigma} d \prod_{\rho,\sigma} , \quad \bar{\partial} = \prod_{\rho,\sigma+1} d \prod_{\rho,\sigma} ,$$

$$\vartheta = \prod_{\rho,\sigma-1} \delta \prod_{\rho,\sigma} , \quad \bar{\vartheta} = \prod_{\rho-1,\sigma} \delta \prod_{\rho,\sigma} .$$

Then

$$d \prod_{\rho,\sigma} = \partial + \bar{\partial}, \quad \delta \prod_{\rho,\sigma} = \vartheta + \bar{\vartheta} .$$

The Laplace-Beltrami operator Δ is defined by the formula $\Delta = \delta d + d\delta$ and its complex analogues are

$$\Box = \bar{\vartheta}\partial + \partial\bar{\vartheta}, \quad \bar{\Box} = \vartheta\bar{\partial} + \bar{\partial}\vartheta .$$

Finally, if $p \geq 2$, let

$$(\Lambda\varphi)_{i_1 \ldots i_{p-2}} = -h^{(ij)} \varphi_{(ij)i_1 \ldots i_{p-2}} ,$$

while, if $p = 0$ or 1, set $\Lambda\varphi = 0$ (Hodge-Weil operator).

The following identities are valid:

$$* \prod_{\rho,\sigma} = \prod_{k-\sigma,k-\rho} *, \quad *\partial = (-1)^{p+1} \vartheta *, \quad *\vartheta = (-1)^p \partial * ;$$

$$\partial^2 = \vartheta^2 = 0, \quad \partial\bar{\partial} + \bar{\partial}\partial = 0, \quad \vartheta\bar{\vartheta} + \bar{\vartheta}\vartheta = 0, \quad \vartheta\partial + \partial\vartheta = 0;$$

$$\Delta \prod_{\rho,\sigma} = \prod_{\rho,\sigma} \Delta = \Box + \bar{\Box}, \quad \Box = \bar{\Box} = \frac{1}{2} \prod_{\rho,\sigma} \Delta ;$$

$$\Lambda\partial - \partial\Lambda = - \sqrt{-1}\, \vartheta .$$

A compact subdomain B of M will be called a finite submanifold if each boundary point q of B has a neighborhood U_q in M in which real coordinates $u^1, u^2, \ldots, u^{2k}$ exist satisfying the following conditions:

i) each u^i is a function of the z^j of class C^∞ and the Jacobian $\partial(u^1, \ldots, u^{2k})/\partial(z^1, \ldots, z^{2k})$ does not vanish in U_q;

ii) U_q is mapped topologically onto a subdomain in the u-space in such a way that the hyperplane $u^{2k} = 0$ corresponds to points of the boundary of B, the coordinates $u^1, \ldots, u^{2k-1}$ being local boundary parameters;

iii) the u^{2k}-curve is orthogonal to the hyperplane $u^{2k} = 0$ and hence g_{ij}, expressed in terms of the coordinates u^i, satisfies on the boundary the condition that $g_{i,2k} = g^{i,2k} = 0$ for $i = 1,2,\ldots,2k-1$. The coordinates u^i will be called boundary coordinates.

A finite manifold is either a finite submanifold with boundary or is a compact (closed) manifold.

The topological boundary operator will be denoted by b. In U_q, $q \in bB$, we set

$$t\varphi = \sum_{i_1<\ldots<i_p<2k} \varphi_{i_1\ldots i_p}\, du^{i_1} \wedge \ldots \wedge du^{i_p},$$

$$n\varphi = \sum_{i_1<\ldots<i_p=2k} \varphi_{i_1\ldots i_p}\, du^{i_1} \wedge \ldots \wedge du^{i_p}.$$

Then, in U_q, $\varphi = t\varphi + n\varphi$, but this decomposition will in general have geometrical meaning only on the boundary bB itself. However, if we choose u^{2k} to be geodesic distance from bB, $u^{2k} > 0$ in $U_q \cap B$, then $t\varphi$ and $n\varphi$ are well-determined p-forms in a sufficiently small boundary strip $0 < u^{2k} < \varepsilon$ covered by the coordinate systems $u^1, \ldots, u^{2k}$. We verify that $*t = n*$, $t* = *n$.

The scalar product of two p-forms φ and ψ on a subdomain D of M is defined to be

$$(\varphi, \psi) = (\varphi, \psi)_D = \int_D \varphi \wedge *\psi,$$

and the corresponding norm is $\|\varphi\| = \sqrt{(\varphi, \varphi)}$. We plainly have

$$(\prod_{\rho,\sigma} \varphi, \psi) = (\varphi, \prod_{\rho,\sigma} \psi);$$

that is, $\prod$ is a self-adjoint operator.

If B is a finite manifold, various Green's identities involving either real or complex operators are valid for all sufficiently regular forms; for example,

$$(\partial\varphi, \partial\psi) - (\varphi, \vartheta\partial\psi) = \int_{bB} \varphi \wedge *(\partial\psi)^{-},$$

$$(\partial\vartheta\varphi, \psi) - (\vartheta\varphi, \vartheta\psi) = \int_{bB} \vartheta\varphi \wedge * \bar{\psi} .$$

In the following sections of this paper all relations which do not essentially involve the complex structure are valid for real Riemannian manifolds of arbitrary dimension m. In case the complex structure is not involved, we use real coordinates x^i or y^i, and denote a point with coordinates x^i by x, one with coordinates y^i by y. If it is desirable to emphasize the presence of complex structure, we use self-conjugate coordinates z^i or ζ^i, and denote a point with coordinates z^i by z, a point with coordinates ζ^i by ζ. We assume, for simplicity, that the manifolds are orientable; this hypothesis is, of course, automatically satisfied in the complex sense.

2. CURRENTS

Let D be an arbitrary subdomain of M (which may coincide with M), and let $\mathcal{X}_p$ denote the space of p-forms φ which are of class C^∞ in D and have compact carriers relative to D. A p-current $T[\varphi]$ on D in the sense of de Rham [10] is a linear functional over the space $\mathcal{X}_{m-p}$ which satisfies the following continuity restriction: for an arbitrary sequence of forms φ_μ, $\varphi_\mu \in \mathcal{X}_{m-p}$, whose carriers are contained in a fixed compact subset K of D, where K is covered by a single coordinate system x^i, $T[\varphi_\mu] \to 0$ $(\mu \to \infty)$ if φ_μ and each partial derivative tend uniformly to zero.

LEMMA 2.1. (Partition of unity). Given a locally finite open covering $\{U_i\}$ of D, there exists a corresponding set of scalars φ_j such that:

(i) $\sum_j \varphi_j = 1$;

(ii) $\varphi_j \in C^\infty$, $0 \leq \varphi_j \leq 1$ everywhere, and the carrier of φ_j is contained in one of the open sets U_i.

This lemma is stated in [10].

Given an (m-p)-form φ of class C^∞ with a non-compact carrier, we say that $T[\varphi]$ is convergent and that

$$T[\varphi] = \sum_i T[\varphi_i \varphi]$$

if the series is convergent for each partition of unity.

The exterior product $T \wedge \psi$ of a p-current T with a q-form $\psi \in C^\infty$ is defined to be

$$T \wedge \psi[\varphi] = T[\psi \wedge \varphi].$$

Furthermore we set

$$dT[\varphi] = (-1)^{p+1} T[d\varphi], \quad \delta T[\varphi] = (-1)^p T[\delta\varphi],$$

$$*T[\varphi] = (-1)^p T[*\varphi], \quad (T, \varphi) = T[*\varphi].$$

We say that T vanishes at a point q of D if there is a neighborhood U_q of q such that $T[\varphi] = 0$ for all φ whose carriers lie in U_q. The carrier of T is the set of all points $q \in D$ where T does not vanish. A p-current is said to be regular at a point q of D if T coincides with a p-form of class C^∞ in some neighborhood of q; otherwise T is called singular at q. The set of singular points of T is called the singular set of T.

Given two p-currents S and T whose singular sets do not meet, we can split S and T such that $S = S'' + s''$, $T = T' + T''$ where S'' and T'' are everywhere regular while the carriers of S' and T' do not meet. In this case we define

$$(S,T) = (S',T'') + (S'',T') + (S'',T'')$$

provided the scalar products on the right converge.

THEOREM 2.1. If T is a p-current in $D \subseteq M$ and if ΔT is regular at the point $q \in D$, then T is also regular at q. Moreover, if dT and δT are both regular at q, T is regular at q.

A proof of this theorem is to be found in [10].

An arbitrary p-current T may be represented as a formal differential form

$$T = T_{(i_1 \ldots i_p)} \, dz^{i_1} \wedge \ldots \wedge dz^{i_p}$$

where the coefficients $T_{i_1\ldots i_p}$ are distributions in the space of local self-conjugate coordinates z^i. Therefore the operator $\prod_{\rho,\sigma}$ may be applied to T:

$$\prod_{\rho,\sigma} T[\varphi] = T[\prod_{k-\sigma,k-\rho} \varphi].$$

We define

$$\partial T[\varphi] = (-1)^{p+1}\, T[\bar{\partial}\varphi], \quad \vartheta T[\varphi] = (-1)^p T[\bar{\vartheta}\varphi] .$$

A p-current T is harmonic if $\Delta T[\varphi] = T[\Delta\varphi] = 0$. By Theorem 2.1 we see that a harmonic current is equal to a harmonic form. Similarly, if $dT[\varphi] = \delta T[\varphi] = 0$, T is equal to a harmonic field. If the p-current T is of type (ρ,σ) and satisfies $\partial T = \vartheta T = 0$ (or $\bar{\partial} T = \bar{\vartheta} T = 0$), it is equal to a complex field.

If T is of type $(\rho, 0)$, then $\bar{\vartheta} T = 0$ automatically and the relation $\bar{\partial} T = 0$ implies that T is holomorphic. Similarly, a current T of type $(0,\sigma)$ satisfying $\partial T = 0$ will be called conjugate holomorphic.

3. GREEN'S OPERATORS ON REAL RIEMANNIAN MANIFOLDS

In this section we make no use of the complex structure, and we therefore assume that M is an orientable Riemannian manifold of dimension m and of class C^∞. Let x,y denote points of M expressed in terms of real local coordinates x^i, y^i respectively.

Given a point y of M, we denote by Y the p-current on M with carrier y whose components are

$$Y_{i_1\ldots i_p} = g_{i_1\ldots i_p,(j_1\ldots j_p)}\, dy^{j_1} \wedge \ldots \wedge dy^{j_p} \tag{3.1}$$

and which satisfies

$$(\varphi, Y) = \varphi(y) \tag{3.2}$$

for every p-form φ. If $p = 0$, Y is the 0-current which has the value 1 at y and is zero everywhere else.

We denote the spaces of norm finite harmonic p-forms and harmonic p-fields by H and F respectively, and we also use H and F to denote projection onto the corresponding spaces.

Let $\Phi = \Phi_p$ denote the space of p-forms φ, $\varphi \in C^\infty$, and let $S = S_p$ be a subspace of Φ. We form a closed space $[S]$ from S by taking the closure in the sense of the norm of all norm-finite p-forms in

S. The space obtained from S by applying the operator d to each of its elements will be denoted by dS, and the spaces δS, ΔS are defined in a similar manner.

As in Section 2, let χ_p be the subspace of Φ_p composed of p-forms whose carriers are compact. Then we have the following formulas of orthogonal decomposition:

$$[\Phi_p] = [\Delta\chi_p] + \mathsf{H}, \tag{3.3}$$

$$[\Phi_p] = [d\chi_{p-1}] + [\delta\chi_{p+1}] + \mathsf{F}. \tag{3.4}$$

We define the subspace E of F to be the orthogonal complement of the spaces $[d\Phi_{p-1}]$ and $[\delta\Phi_{p+1}]$, that is,

$$[\Phi_p] = [d\Phi_{p-1}] \cup [\delta\Phi_{p+1}] + \mathsf{E}. \tag{3.5}$$

It is obvious that $\mathsf{E} \subseteq \mathsf{H}$ and that $\mathsf{E} = \mathsf{H} = \mathsf{F}$ if M is closed. Moreover, the following result is easily proved:

LEMMA 3.1. If B is a finite manifold, $B \subset M$, E may be regarded as the subspace of F composed of harmonic p-forms on M which vanish identically outside B.

Let $\mathcal{C} = \mathcal{C}_p$ be the class of p-currents T on M which converge for every p-form φ, $\varphi \in \Phi_p$, $\|\varphi\| < \infty$, and let K denote an arbitrary closed subspace of H. If $T \in \mathcal{C}_p$, we define

$$\mathsf{K}T[\varphi] = T[\mathsf{K}\varphi]. \tag{3.6}$$

Since the p-current Y defined above plainly belongs to the class $\mathcal{C}_p$, $\mathsf{K}Y$ exists and is a symmetric double form $k(x,y)$ of degree p in x and in y. The double form k is the kernel of the projection operator K (reproducing kernel for K). Given a p-form φ, let

$$|\varphi| = \sqrt{\varphi_{(i_1 \ldots i_p)}\,\varphi^{(i_1 \ldots i_p)}}\,. \tag{3.7}$$

By applying the Schwartz inequality to the relation

$$\varphi(y) = (\varphi, \mathsf{K}Y) = (\varphi, k), \quad \varphi \in \mathsf{K},$$

we see that there exists a positive number $K(y)$ depending only on the point y such that

$$|\varphi(y)| \leq K(y)\,\|\varphi\| \tag{3.8}$$

for every $\varphi \in \mathcal{K}$. The inequality (3.8) shows that $\mathcal{K}$ is a Hilbert space in which convergence in the sense of the norm implies uniform point-wise convergence over any compact subdomain of M. The important spaces are the full space $\mathcal{H}$ and the subspaces $\mathcal{F}$ and $\mathcal{E}$.

If there exists a p-current $G = G_{\mathcal{K}}$ on M which satisfies

$$\Delta G = Y - \mathcal{K}Y \tag{3.9}$$

for each $y \in M$, we say that M possesses a Green's operator for the class $\mathcal{K}$. If G exists, then by Theorem 2.1 G has a kernel $g(x,y) = g_{\mathcal{K}}(x,y)$ which is a double form of degree p in x and in y. The kernel g is obviously determined only up to an arbitrary p-form harmonic in x. It satisfies $\Delta_x g = -k(x,y)$, $x \neq y$; for fixed y, g has a singularity at y where the principal term is

$$(3.10)\quad \frac{1}{(m-2)s_m r^{m-2}}\, g_{(i_1 \dots i_p),(j_1 \dots j_p)} dx^{i_1} \wedge \dots \wedge dx^{i_p} \cdot dy^{j_1} \wedge \dots \wedge dy^{j_p}.$$

Here m is the dimension of M, s_m is the volume of the unit $(m-1)$-sphere in euclidean space, and $r = r(x,y)$ is the geodesic distance between the points x and y. We remark that the difference of any two Green's kernels is regular at y; this follows from Theorem 2.1.

We set

$$G\varphi = (\varphi(x), g(y,x)), \quad \mathcal{K}\varphi = (\varphi(x), k(y,x));$$

then (3.9) may be written in the form

$$(3.9)' \quad \Delta G\varphi = \varphi(y) - \mathcal{K}\varphi(y).$$

Similarly,

$$GY = (Y,G) = g(x,y).$$

Now let M_1, M_2 be Riemannian manifolds and suppose that $M_1 \subseteq M_2$. In other words, let M_1 be a submanifold of M_2. We write $\mathcal{K}_2 \subseteq \mathcal{K}_1$ if every φ of $\mathcal{K}_2$ when restricted to M_1 belongs to $\mathcal{K}_1$. We define the kernel $k_1(x,y)$ of $\mathcal{K}_1$ to be zero if either x or y lies outside M_1. It is clear that

$$(3.11) \quad \mathcal{H}_2 \subseteq \mathcal{H}_1, \quad \mathcal{F}_2 \subseteq \mathcal{F}_1.$$

On the other hand, if M_1 is a finite submanifold of M_2, then by Lemma 3.1 every $\varphi \in \mathcal{E}_1$ can be made into an element of $\mathcal{E}_2$ merely be defining it to be identically zero in the difference domain $M_2 - M_1$. Hence we have:

LEMMA 3.2. If $M_1 \subseteq M_2$, then $E_1 \subseteq E_2$.

The following simple theorem is a statement concerning one Riemannian manifold imbedded into another:

THEOREM 3.1. If $M_1 \subseteq M_2$, $K_1 \supseteq K_2$, and if a Green's operator G_2 for the class K_2 exists on M_2, then

$$G_1 = G_2 - G_2 K_1 \tag{3.12}$$

is a Green's operator on M_1 for the class K_1.

In particular, if M_1 coincides with M_2 and if $K_1 \supseteq K_2$, then G_1 exists if G_2 exists.

To prove the theorem, let $y \in M_1$ and let $Q = Y - K_1 Y$. Then

$$K_2 Q = K_2 Y - K_2 K_1 Y = K_2 Y - K_1 K_2 Y = K_2 Y - K_2 Y = 0,$$

and hence by (3.9)

$$\Delta G_2 Q = Q = Y - K_1 Y.$$

Thus $G_2 Q$ is a Green's operator on M_1 for the class K_1.

Since the E-classes do not satisfy the inclusion relationship required by Theorem 3.1, we need another theorem for these classes, namely:

THEOREM 3.2. If M_1 is a finite submanifold of M_2 and if G_2 exists on M_2 for the class K_2, then G_1 exists on M_1 for the class $K_1 = E_1 \cap K_2$.

Let the p-forms of K_2 be orthogonalized over M_2, and let the p-forms of $K_2 - E_1$ be denoted by φ_i, those of $E_1 \cap K_2$ by ψ_i. Since the φ_i are linearly independent with respect to constants over $M_2 - M_1$, we assume that the φ_i are orthonormalized over $M_2 - M_1$. Similarly we may suppose that the ψ_i are orthonormalized. Let $y \in M_1$ and set

$$\Phi Y = \begin{cases} \Sigma \varphi_i(x)\varphi_i(y), & x \in M_2 - M_1, \\ 0, & x \in M_1, \end{cases}$$

$$\Psi Y = K_1 Y = \Sigma \psi_i(x)\, \psi_i(y).$$

Then $Q = Y - \Phi Y - \Psi Y$ obviously satisfies $K_2 Q = 0$ and hence, by (3.9),

$$\Delta G_2 Q = Q = Y - \Phi Y - \Psi Y.$$

Since ΦY vanishes for $x \in M_1$, $G_2 Q$ is a Green's operator on M_1 for the class $K_1 = E_1 \cap K_2$.

We denote the Green's operators which correspond to the classes H, E and F by G, G_E and N respectively. If E vanishes, we call G_E a fundamental operator.

4. GREEN'S OPERATORS ON FINITE REAL RIEMANNIAN MANIFOLDS

Let B denote an arbitrary finite Riemannian manifold. Then:

LEMMA 4.1. B can be imbedded isometrically in a closed Riemannian manifold R of the same dimension and of class C^∞.

A proof, which is easily supplied, will be omitted

LEMMA 4.2. (de Rham [10]). An operator G for the class H exists on every closed manifold.

Combining Theorems 3.1 and 3.2 with Lemmas 4.1 and 4.2, we obtain:

THEOREM 4.1. Operators G, G_E and N exist on every finite manifold.

However, the operators G, G_E and N are not unique and are determined only up to an additive harmonic p-form in x. But we can make them unique by fixing their boundary values in a suitable manner, and for this we require the following lemmas concerning boundary-values:

LEMMA 4.3. (First boundary-value problem for harmonic p-forms [4]). Let θ be a continuous p-form defined in a boundary neighborhood of the finite manifold B. Then there exists a harmonic p-form φ in B with $t\varphi = t\theta$, $n\varphi = n\theta$ on the boundary.

LEMMA 4.4. (Second boundary-value problem for harmonic p-forms [4]). Let θ be a p-form defined in a boundary neighborhood of the finite manifold B which is continuous together with $d\theta$, $\delta\theta$ in this neighborhood and satisfies

$$\int_{bB} [\psi \wedge *d\theta - \delta\theta \wedge *\psi] = 0 \tag{4.1}$$

for every harmonic p-field ψ in B which is continuous in the closure of B. Then there exists a harmonic p-form φ in B with $t\delta\varphi = t\delta\theta$, $nd\varphi = nd\theta$ on the boundary.

In Lemmas 4.3 and 4.4 the harmonic p-form φ is unique if and only if $\mathsf{E} = 0$.

LEMMA 4.5. ([5a,b], [7b]). Let θ satisfy the conditions of Lemma 4.3. Then there exists a p-form φ which satisfies $d\delta\varphi = 0$ in B with $n\varphi = n\theta$ on the boundary. If $\theta = d\tau$ in a boundary neighborhood, there is a unique (p-1)-form ψ which is orthogonal to closed forms in B and satisfies $\Delta d\psi = 0$ with $nd\psi = nd\tau$ on the boundary.

Lemma 4.3 and 4.4 are self-dual under the operation $*$ while the dual of Lemma 4.5 is obtained by interchanging d, δ and also n, t.

The proofs of Lemmas 4.3-4.5 can be based on a generalization of the Poincaré-Fredholm method for integral equations combined with the method of orthogonal projection. In the Poincaré-Fredholm method the discontinuity of certain singular integrals across the boundary of the domain gives rise to integral equations whose solutions must then be shown to exist. Here we discuss only the discontinuity behavior of the integrals.

By Lemma 4.1, B can be imbedded in a closed R. By Lemma 4.2 the closed R has a Green's operator $G = G_R$ which can be made unique by the requirement that $\mathsf{H}G = 0$. Then G is self-adjoint and

$$dG = Gd, \quad \delta G = G\delta \ . \tag{4.2}$$

The corresponding kernel $g_p(x,y)$ therefore satisfies the relations

$$g_p(x,y) = g_p(y,x); \quad d_x g_p(x,y) = \delta_y g_{p+1}(x,y). \tag{4.3}$$

Given a p-form φ which is continuous in the closure of B, we write

$$\lambda_{p+1} = \int_{bB} \varphi \wedge *g_{p+1}, \quad \eta_{p-1} = \int_{bB} g_{p-1} \wedge *\varphi \ . \tag{4.4}$$

LEMMA 4.6. The expressions $\delta\lambda_{p+1}$, $d\eta_{p-1}$ decrease by the amounts $t\varphi$, $n\varphi$ respectively as the boundary is crossed from the interior to the exterior of B. In particular, $n\delta\lambda_{p+1}$ and $td\eta_{p-1}$ are continuous across the boundary.

This lemma may be proved as follows. Let $B\varphi$ be equal to φ in B, equal to zero outside B. By Green's formula applied to B,

$$B\varphi = \delta\lambda_{p+1} + d\eta_{p-1} + I \tag{4.5}$$

where

$$I = (d\varphi, dg_p) + (\delta\varphi, \delta g_p) + (\varphi, k).$$

Here $k = k_H$ is the kernel of the operator H for R, $B \subset R$, and we see that I is continuous across the boundary of B.

Each point $p \in bB$ has a neighborhood in which boundary coordinates $u^1, \ldots, u^m$ are valid. If we choose u^m to be geodesic distance from the boundary, $u^m > 0$ in B, then (compare Section 1) $t\varphi$ and $n\varphi$ are well-defined p-forms in a boundary strip S: $0 < u^m < \varepsilon$. If we apply formula (4.5) to each of the forms $t\varphi$, $n\varphi$ in the strip S, we obtain

$$St\varphi = \delta\lambda_{p+1} + J_t, \quad Sn\varphi = d\eta_{p-1} + J_n,$$

where J_t and J_n are continuous across the boundary of B. Thus Lemma 4.6 is proved.

On the basis of Lemma 4.5 (or rather its dual) we define a kernel $\mu(x,y)$ of degree p for B which is characterized by the properties:

$$\begin{gathered}\Delta_x\delta_x\mu = 0, \quad x \neq y; \quad \delta_x(\mu - g) \text{ regular at } y; \\ (\psi(x), \mu(x,y)) = 0 \text{ if } \delta\psi = 0 \text{ in } B; \quad t_x\delta_x\mu = 0 \text{ on the boundary.}\end{gathered} \tag{4.6}$$

The dual kernel which satisfies (4.6) with d, δ and n, t interchanged will be denoted by ν. Let

$$e(x,y) = \Delta_x\mu = d_x\delta_x\mu, \quad f(x,y) = \Delta_x\nu = \delta_x d_x\nu. \tag{4.7}$$

Then e and f are the dual Kodaira kernels (see [9a]). The following lemma is obtained immediately by applying Green's formulas:

LEMMA 4.7. The kernel of F is given by the formula

$$k_F = -e - f \tag{4.8}$$

or, symbolically,

$$Y = e + f + FY. \tag{4.8)'}$$

This formula is the orthogonal decomposition of Y which corresponds to (3.4).

Let g be the kernel of G on a finite B. By Lemma 4.3 there exists a p-form $\psi = \psi(x,y)$, harmonic in $x \in B$, such that $t_x \psi = t_x g$, $n_x \psi = n_x g$ on the boundary. We may therefore assume that $t_x g = n_x g = 0$ on the boundary. It follows from Green's formula that

$$\int_{bB} [\varphi \wedge *dg - \delta g \wedge * \varphi] = 0, \qquad \varphi \in H.$$

Since $t\varphi$ and $n\varphi$ can be chosen arbitrarily on the boundary, we conclude that $t_x \delta_x g = n_x d_x g = 0$ on the boundary. We make $g = g_p$ unique by the requirement that $E g = 0$.

Let φ, ψ be p-forms, and set

$$D(\varphi, \psi) = (d\varphi, d\psi) + (\delta\varphi, \delta\psi). \tag{4.9}$$

Then $D(\varphi, \psi)$ is just the Dirichlet integral for p-forms. Let $n = n_p$ denote the kernel of the operator N. If ψ is a p-form in B, we have

$$D(\psi, n) = (d\psi, dn) + (\delta\psi, \delta n) = (\psi, \Delta n) + \psi + \int_{bB} [\psi \wedge *dn - \delta n \wedge * \psi]$$

$$= \psi - F\psi + \int_{bB} [\psi \wedge *dn - \delta n \wedge * \psi]$$

since $\Delta_x n = -k_F$, the kernel of F. Hence, if ψ is a harmonic p-field,

$$\int_{bB} [\psi \wedge *dn - \delta n \wedge * \psi] = 0,$$

and therefore n satisfies (4.1). It follows that we can find a p-form $\varphi = \varphi(x,y)$, harmonic in x, such that $t_x \delta_x \varphi = t_x \delta_x n$, $n_x d_x \varphi = n_x d_x n$ on the boundary. Thus we may assume that $t_x \delta_x n = n_x d_x n = 0$ on the boundary. We make $n = n_p$ unique by the requirement that $F n = 0$. It then follows readily from Green's formulas that n is a symmetric kernel: $n(x,y) = n(y,x)$. For a p-form φ we have the decomposition formula

$$\varphi = \big(\Delta\varphi(x), n(x,y)\big) + F\varphi + \int_{bB} [n \wedge *d\varphi - \delta\varphi \wedge *n].$$

If we let

$$P_n \varphi = \int_{bB} [n \wedge *d\varphi - \delta\varphi \wedge *n], \tag{4.10}$$

then

$$N\Delta\varphi = \varphi - F\varphi - P_n\varphi \tag{4.11}$$

where $\Delta P_n\varphi$ is a harmonic p-field and $t\delta P_n\varphi = t\delta\varphi$, $ndP_n\varphi = nd\varphi$ on the boundary. If μ and ν are the kernels introduced above, it is clear that the sum $\mu + \nu$ has all the properties of n. Since n is unique, we conclude that

$$n = \mu + \nu . \tag{4.12}$$

Finally, the kernel g_E of G_E may be normalized in two ways. By subtraction of a suitable p-form, harmonic in x, we may suppose that $t_x g_E = n_x g_E = 0$ on the boundary. The kernel which satisfies this condition and which is orthogonal to the space E will be denoted by $\mathfrak{g} = \mathfrak{g}(x,y)$, the corresponding Green's operator by $\mathcal{G}$. The kernel $\mathfrak{g}$ is symmetric, and we have the decomposition

$$\varphi = (\Delta\varphi(x), \mathfrak{g}(x,y)) + E\varphi - \int_{bB} [\varphi \wedge *d\mathfrak{g} - \delta\mathfrak{g} \wedge *\varphi].$$

If we write

$$P_{\mathfrak{g}}\varphi = - \int_{bB} [\varphi \wedge *d\mathfrak{g} - \delta\mathfrak{g} \wedge *\varphi], \tag{4.13}$$

then

$$\mathcal{G}\Delta\varphi = \varphi - E\varphi - P_{\mathfrak{g}}\varphi \tag{4.14}$$

where $P_{\mathfrak{g}}\varphi$ is a harmonic p-form satisfying $tP_{\mathfrak{g}}\varphi = t\varphi$, $nP_{\mathfrak{g}}\varphi = n\varphi$ on the boundary. Thus $P_{\mathfrak{g}}$ signifies projection onto solutions of the first boundary-value problem.

The kernel g of G which satisfies $t_x g = n_x g = t_x\delta_x g = n_x d_x g = 0$ on the boundary is not symmetric, and therefore G is not a self-adjoint operator. We therefore write

$$G\varphi = (\varphi(x), g(y,x)), \quad G'\varphi = (\varphi(x), g(x,y)). \tag{4.15}$$

Since the kernels $n, \mathfrak{g}$ are symmetric, the corresponding operators N, $\mathcal{G}$ are self-adjoint. We summarize the above results in the following theorem:

THEOREM 4.2. On a finite manifold B there are unique operators G, N, $\mathcal{G}$ satisfying

$$\begin{aligned} \Delta G\varphi &= \varphi - H\varphi, & G'\Delta\varphi &= \varphi - H\varphi, & EG &= 0; \\ \Delta N\varphi &= \varphi - F\varphi, & N\Delta\varphi &= \varphi - F\varphi - P_n\varphi, & FN &= 0; \\ \Delta \mathcal{G}\varphi &= \varphi - E\varphi, & \mathcal{G}\Delta\varphi &= \varphi - E\varphi - P_{\mathfrak{g}}\varphi, & E\mathcal{G} &= 0. \end{aligned} \tag{4.16}$$

Here

$$d\Delta P_n\varphi = \delta\Delta P_n\varphi = 0, \quad \Delta P_{\mathfrak{g}}\varphi = 0, \tag{4.17}$$

and on the boundary,

$$t\delta P_n\varphi = t\delta\varphi, \; ndP_n\varphi = nd\varphi; \quad tP_{\mathfrak{g}}\varphi = t\varphi, \; nP_{\mathfrak{g}}\varphi = n\varphi. \tag{4.18}$$

If T is a p-current of class $\mathcal{C}$ on B and if $G = G_K$ is a Green's operator, we define GT, $G'T$ by the formulas

$$(GT,\varphi) = (T,G'\varphi); \quad (G'T,\varphi) = (T,G\varphi) \tag{4.19}$$

where G' is the adjoint operator of G. Then

$$(\Delta GT,\varphi) = (GT,\Delta\varphi) = (T,G'\Delta\varphi). \tag{4.20}$$

In the case $G = G_H$, we therefore have

$$\Delta GT = T - HT, \quad G'\Delta = \Delta G. \tag{4.21}$$

If φ has a compact carrier, $\varphi \in \mathfrak{X}_p$, then

$$N\Delta\varphi = \varphi - F\varphi, \quad \mathcal{G}\Delta\varphi = \varphi - E\varphi,$$

by (4.11), (4.14). Thus

$$\Delta NT = T - FT, \quad \Delta\mathcal{G}T = T - ET. \tag{4.22}$$

As an example, let T be a bounding (m-p)-cycle C^{m-p} whose carrier is contained in B. Then NC^{m-p} is a p-current and, since $F C^{m-p} = 0$,

$$\Delta NC^{m-p} = C^{m-p}.$$

Thus NC^{m-p} is harmonic except on C^{m-p}. Plainly

$$NC^{m-p} = *_y \int_{C^{m-p}} n_{m-p}(x,y), \tag{4.23}$$

and therefore

(4.24) $$t_y \delta_y N C^{m-p} = n_y d_y N C^{m-p} = 0$$

on the boundary. In the special case $p = m$ we may take $C^{m-p} = C^0 = y_1 - y_2$ where y_1, y_2 are arbitrary points in B. Then

(4.25) $$\Delta N C^0 = Y_1 - Y_2,$$

and $*NC^0$ is the harmonic function on B which has the typical singularities at y_1, y_2 and has a vanishing normal derivative on the boundary. In the case of scalars (functions), E is obviously empty, so $*NC^0$ is the unique harmonic function with these properties.

As a further example, let y be a point of B and take T to be the (p+1)-current dY. Then $\mathsf{F} T = 0$, $\Delta NT = T = dY$, so NT is a harmonic (p+1)-form whose singularity at y is the same as that of $d_x g(x,y)$ where $g = g_p$ is the kernel of a Green's operator G of degree p. Moreover $t\delta NT = ndNT = 0$ on the boundary.

5. GREEN'S OPERATORS ON KÄHLER MANIFOLDS

Now assume that M is a Kähler manifold of complex dimension k with a Kähler metric of class C^∞, and let z, ζ denote points of M expressed in terms of local self-conjugate coordinates z^i, ζ^i respectively. Points of M may also be expressed in terms of real coordinates, and then x, y are points of M expressed in terms of real local coordinates x^i, y^i as in Section 3.

We restrict ourselves to the space of p-forms φ of type (ρ,σ). Given a point y of M, we consider the p-current whose components are

(5.1) $$(\prod_{\rho,\sigma} Y)_{i_1 \ldots i_p} = \prod_{\rho,\sigma} {}_{i_1 \ldots i_p, (j_1 \ldots j_p)} \, dy^{j_1} \wedge \ldots \wedge dy^{j_p}.$$

For a p-form φ of type (ρ,σ) we then have (3.2). The spaces H and E are to be replaced by the corresponding spaces $\prod_{\rho,\sigma}\mathsf{H}$, $\prod_{\rho,\sigma}\mathsf{E}$ obtained by applying the operator $\prod_{\rho,\sigma}$ to each of their elements. To simplify notation, we drop the operator $\prod$ and understand, for example that $\mathsf{H} = \prod_{\rho,\sigma} \mathsf{H}$ is the space of norm-finite harmonic p-forms of type (ρ,σ) (and similarly for E). On the other hand, the complex analogues of the space F are the complex fields φ of type (ρ,σ) which satisfy $\bar\partial\varphi = \vartheta\varphi = 0$ or $\partial\varphi = \bar\vartheta\varphi = 0$. The space of norm-finite p-fields of type (ρ,σ) satisfying $\bar\partial\varphi = \vartheta\varphi = 0$ will be denoted by F, the space of those of type (ρ,σ) satisfying $\partial\varphi = \bar\vartheta\varphi = 0$ by $\bar{\mathsf{F}}$. The intersection $\mathsf{F} \cap \bar{\mathsf{F}}$ is then just the space of norm-finite p-forms φ of type (ρ,σ) satisfying $d\varphi = \delta\varphi = 0$. As in Section 3, H, E, F, $\bar{\mathsf{F}}$ will also be used to denote projection onto the corresponding spaces. With these conventions, the orthogonal decomposition formulas (3.3) and (3.5) remain unchanged for the spaces of type (ρ,σ) while formula (3.4) is replaced by

$$[\Phi_p] = [\bar{\partial}\chi_{p-1}] + [\vartheta\chi_{p+1}] + \mathsf{F} = [\partial\chi_{p-1}] + [\bar{\vartheta}\chi_{p+1}] + \bar{\mathsf{F}}. \qquad (5.2)$$

If M is closed, it is clear that $\mathsf{H} = \mathsf{E} = \mathsf{F} = \bar{\mathsf{F}}$. The analogue of Lemma 3.1 is

LEMMA 5.1. If B is a finite Kähler manifold, $B \subset M$, E may be regarded as the subspace of $\mathsf{F} \cap \bar{\mathsf{F}}$ composed of harmonic p-forms on M which vanish identically outside B.

If either ρ or σ has one of the values 0 or k, $\mathsf{F} \cap \bar{\mathsf{F}}$ is composed of forms which are a) holomorphic $(\sigma = 0)$; b) become holomorphic by taking their conjugates or applying the operator $*$ or both. Therefore we conclude from analytic continuation that, if ρ or σ is equal to 0 or k, E is empty. Hence:

LEMMA 5.2. If B is a finite Kähler manifold, the space E is empty whenever ρ or σ has one of the values 0, k.

The remaining considerations of Section 3 go over without essential change, it being understood that p-currents are of type (ρ,σ), in particular that $Y = \prod_{\rho,\sigma} Y$. Symmetric double forms become Hermitian; for example, the kernel k of a projection operator $\mathsf{K} = \prod_{\rho,\sigma}\mathsf{K}$ is Hermitian. Moreover, in formulas such as (3.10) it is clear that $g_{i_1\ldots i_p, j_1\ldots j_p}$ is to be replaced by $\prod_{\rho,\sigma} {}_{i_1\ldots i_p, j_1\ldots j_p}$. Finally if g is the kernel of a Green's operator G we have

$$G'\varphi = \prod_{\rho,\sigma}{}_y G'\varphi = (\varphi(x), g(x,y)) = (\varphi(x), \prod_{\rho,\sigma}{}_x \prod_{\sigma,\rho}{}_y g(x,y)),$$

and therefore

$$g(x,y) = \prod_{\rho,\sigma}{}_x \prod_{\sigma,\rho}{}_y g(x,y) \qquad (5.3)$$

or, in terms of self-conjugate coordinates,

$$g(z,\zeta) = \prod_{\rho,\sigma}{}_z \bar{\prod_{\rho,\sigma}}{}_\zeta g(z,\bar{\zeta}). \qquad (5.3)'$$

Here we place a bar over ζ as a reminder of this property. Formula (5.3) is equivalent to

$$(5.4) \qquad \prod_{\rho,\sigma} G = G \prod_{\rho,\sigma} .$$

6. GREEN'S OPERATORS ON FINITE KÄHLER MANIFOLDS

If B is a finite Kähler manifold, the Green's operators G and $G_{\mathfrak{E}}$ are obtained from the corresponding real Green's operators merely by applying the projection operator $\prod_{\rho,\sigma}$. On the other hand, the operator N does not stand in such a simple relation to the corresponding real operator. Moreover, since a finite Kähler manifold cannot generally be imbedded in a closed Kähler manifold, the reasoning by which we proved the existence of N in Section 4 will apply only if we assume that B is a submanifold of a closed Kähler manifold. Although it is possible to establish the existence of N without this assumption, we suppose for simplicity throughout this section that B is a submanifold of a closed Kähler manifold and then Theorem 4.1 is valid by the same reasoning as that used in the real case.

The boundary-value statements of Lemmas 4.3 - 4.5 also remain valid for forms of type (ρ,σ) with the real operators d, δ replaced by the complex operators $\bar{\partial}, \vartheta$ (or $\partial, \bar{\vartheta}$). In fact, the same Poincaré-Fredholm method for integral equations can be applied once Lemma 4.6 has been shown to be valid for the complex operators and this can be demonstrated.

LEMMA 6.1. Let φ be of type (ρ,σ). Then the expressions $\vartheta\lambda_{p+1}, \bar{\partial}\eta_{p-1}$ decrease by the amounts $t\varphi$, $n\varphi$ respectively as the boundary is crossed from the interior to the exterior of B. In particular, $n\vartheta\lambda_{p+1}$ and $t\bar{\partial}\eta_{p-1}$ are continuous across the boundary.

Since

$$\varphi = \prod_{\rho,\sigma} \varphi ,$$

we have by (4.5)

$$B\varphi = \prod_{\rho,\sigma} B\varphi = (\vartheta + \bar{\vartheta})\lambda_{p+1} + (\partial + \bar{\partial})\eta_{p-1} + I$$

and Lemma 6.1 will follow if we can show that the differences $(\vartheta - \bar{\vartheta})\lambda_{p+1}$, $(\partial - \bar{\partial})\eta_{p-1}$ are continuous across the boundary.

Let U be a neighborhood of the singular point ζ of g_p. Then, as U shrinks to the singular point ζ, the integrals

$$\int_{bU} ((\vartheta - \bar{\vartheta})g_p)^- \wedge *\varphi, \quad \int_{bU} \varphi \wedge *((\partial - \bar{\partial})g_p)$$

tend to zero. Since only the asymptotic behavior is involved, we may suppose that U is a euclidean domain and that

$$g_p(z, \bar{\zeta}) = \frac{1}{2(k-1)s_k r^{2(k-1)}} \prod_{\rho,\sigma}(i_1 \ldots i_p),(j_1 \ldots j_p) \, dz^{i_1} \wedge \ldots \wedge dz^{i_p} \cdot d\zeta^{j_1} \wedge \ldots \wedge d\zeta^{j_p}.$$

If ζ is chosen as the origin of the coordinates z^i, we verify that

$$(\vartheta_z g_p)^- = \frac{1}{s_k r^{2k}} \prod_{\rho,\sigma}(j_1 \ldots j_p),\alpha(i_1 \ldots i_{p-1}) \, z^\alpha \, dz^{i_1} \wedge \ldots \wedge dz^{i_{p-1}} \cdot d\zeta^{j_1} \wedge \ldots \wedge d\zeta^{j_p}$$

where, in the summation, α is restricted to values between 1 and k. Hence

$$(\vartheta_z g_p)^- \wedge *\varphi = \frac{1}{s_k r^{2k}} \, g_{(i_1 \ldots i_{2k-1})}^{(r_1 \ldots r_{p-1})(t_1 \ldots t_{2k-p})} \cdot \prod_{\rho,\sigma}(j_1 \ldots j_p),\alpha(r_1 \ldots r_{p-1}) \, (*\varphi)_{(t_1 \ldots t_{2k-p})} \, z^\alpha \, dz^{i_1} \wedge \ldots \wedge dz^{i_{2k-1}} \cdot d\zeta^{j_1} \wedge \ldots \wedge d\zeta^{j_p}.$$

If in this expression ϑ_z is replaced by $\bar{\vartheta}_z$, then α is replaced by an index $\bar{\alpha}$ whose values range from $k+1$ to $2k$. In the limit we obtain

$$\lim \, (\vartheta_z g_p)^- \wedge *\varphi = \frac{1}{s_k} \, g_{(i_1 \ldots i_{2k-1})}^{(r_1 \ldots r_{p-1})(t_1 \ldots t_{2k-p})} \cdot \prod_{\rho,\sigma}(j_1 \ldots j_p),\alpha(r_1 \ldots r_{p-1}) \, (*\varphi(0))_{(t_1 \ldots t_{2k-p})} \cdot \int_{bU_1} z^\alpha \, dz^{i_1} \wedge \ldots \wedge dz^{i_{2k-1}} \cdot d\zeta^{j_1} \wedge \ldots \wedge d\zeta^{j_p}$$

where U_1 is the sphere of unit radius. By Stokes' theorem

$$\int_{bU_1} z^{\alpha} dz^{i_1} \wedge \ldots \wedge dz^{i_{2k-1}} = \int_{U_1} dz^{\alpha} \wedge dz^{i_1} \wedge \ldots \wedge dz^{i_{2k-1}}.$$

The equality

$$\lim \int_{bU} (\vartheta_z g_p)^- \wedge *\varphi = \lim \int_{bU} (\bar{\vartheta}_z g_p)^- \wedge *\varphi$$

follows since

$$g_{\alpha_1 \ldots \alpha_{\sigma-1} \bar{\beta} \bar{\beta}_1 \ldots \beta_{\rho-1} t_1 \ldots t_{2k-p}}{}^{\alpha_1 \ldots \alpha_{\sigma-1} \bar{\beta} \bar{\beta}_1 \ldots \bar{\beta}_{\rho-1} t_1 \ldots t_{2k-p}}$$

$$\cdot \prod_{\rho,\sigma} j_1 \ldots j_p, \alpha\alpha_1 \ldots \alpha_{\sigma-1} \bar{\beta} \bar{\beta}_1 \ldots \bar{\beta}_{\rho-1} \quad dz^{\alpha} \wedge dz^{\alpha_1} \wedge \ldots \wedge dz^{\alpha_{\sigma-1}} \wedge dz^{\bar{\beta}} \wedge dz^{\bar{\beta}_1} \wedge \ldots \wedge dz^{\bar{\beta}_{\rho-1}} =$$

$$g_{\alpha\alpha_1 \ldots \alpha_{\sigma-1} \bar{\beta}_1 \ldots \bar{\beta}_{\rho-1} t_1 \ldots t_{2k-p}}{}^{\alpha \alpha_1 \ldots \alpha_{\sigma-1} \bar{\beta}_1 \ldots \bar{\beta}_{\rho-1} t_1 \ldots t_{2k-p}}$$

$$\cdot \prod_{\rho,\sigma} j_1 \ldots j_p, \bar{\beta}\alpha\alpha_1 \ldots \alpha_{\sigma-1} \bar{\beta}_1 \ldots \bar{\beta}_{\rho-1} \quad dz^{\bar{\beta}} \wedge dz^{\alpha} \wedge dz^{\alpha_1} \wedge \ldots \wedge dz^{\alpha_{\sigma-1}} \wedge dz^{\bar{\beta}_1} \wedge \ldots \wedge dz^{\bar{\beta}_{\rho-}}$$

where unbarred indices have values between 1 and k, barred indices values between $k+1$ and $2k$.

Finally,

$$(\vartheta\varphi, \vartheta g_p) - (\bar{\vartheta}\varphi, \bar{\vartheta} g_p) = (\varphi, (\bar{\partial}\vartheta - \partial\vartheta) g_p) - \int_{bB} ((\vartheta - \bar{\vartheta}) g_p)^- \wedge *\varphi$$

where

$$\bar{\partial}_z \vartheta_z g_p = \bar{\partial}_z \partial_\zeta g_{p-1}, \quad \partial_z \bar{\vartheta}_z g_p = \partial_z \bar{\partial}_\zeta g_{p-1}.$$

It is readily verified that the difference

$$(\bar{\partial}_z \partial_\zeta - \partial_z \bar{\partial}_\zeta) g_{p-1}$$

is of order $1/r^{2k-1}$ at most, and it follows that $(\partial - \bar{\partial})\eta_{p-1}$ is continuous across the boundary. Similarly, $(\vartheta - \bar{\vartheta})\lambda_{p+1}$ is continuous across the boundary.

As a simple illustration, consider the integral

$$\int_{bU} \varphi \wedge *((\partial - \bar{\partial})g_0)^-$$

in the case $k = 1$ of euclidean space, and take U to be the circle $|z| < \varepsilon$. We have

$$*(\partial_z g_0)^- = \frac{1}{4\pi i}\frac{d\bar{z}}{\bar{z}}, \qquad *(\bar{\partial} g_0) = -\frac{1}{4\pi i}\frac{dz}{z},$$

and therefore on $|z| = \varepsilon$, $z = \varepsilon \exp(\sqrt{-1}\,\theta)$,

$$*(\partial_z g_0)^- = *(\bar{\partial}_z g_0) = -\frac{1}{4\pi}\,d\theta, \qquad *((\partial - \bar{\partial})g_0)^- = 0.$$

For ζ on the boundary of B, we have by the usual reasoning

$$2t\bar{\vartheta}\lambda_{p+1} = -2t\,\vartheta\int_{bB}\varphi \wedge *(g_{p+1})^- + \tfrac{1}{2}t\varphi, \quad 2n\bar{\partial}\eta_{p-1} = 2n\,\partial\int_{bB}(g_{p-1})^- \wedge *\varphi + \tfrac{1}{2}n\varphi$$

where the integrals are to be interpreted in the principal-value sense.

On the basis of the complex analogues of Lemma 4.5 we can define kernels μ and ν satisfying the corresponding conditions in terms of the complex operators $\bar{\partial}, \vartheta$. We thus obtain complex analogues of the Kodaira kernels e, f for which Lemma 4.7 remains valid. In particular, in the case $p = 0$ $(\rho = \sigma = 0)$ we have ([7a,b])

$$k_F = -2\vartheta_z \bar{\partial}_z \nu(z,\bar{\zeta}) \tag{6.1}$$

where k_F is the reproducing kernel of holomorphic functions on B.

Theorem 4.2 remains valid for the complex operators. In particular, for $\varphi \in F$ we have the "Cauchy" formula

$$\varphi = E\varphi - 2\int_{bB}[\varphi \wedge *(\bar{\partial} g)^- - (\vartheta g)^- \wedge *\varphi]. \tag{6.2}$$

If either ρ or σ is equal to 0 or k, $E\varphi = 0$ by Lemma 5.2. In particular, if $\sigma = 0$, then $\vartheta g = 0$ automatically and we obtain simply

$$\varphi = -2 \int_{bB} \varphi \wedge *(\delta_{\gamma})^{-}. \tag{6.2)'}$$

In the case of euclidean space of complex dimension $k = 1$, this formula reduces to the classical formula

$$\varphi(\zeta) = \frac{1}{2\pi i} \int_{bB} \varphi(z) \frac{dz}{z - \zeta}.$$

Finally, the current formulas (4.31) and (4.32) are valid in the complex case. If T is a p-current of type (ρ, σ) satisfying $T = \delta Q$, then $F T = 0$ and we obtain

$$\Delta NT = T. \tag{6.3}$$

Hence NT is harmonic at points which do not lie on the carrier of T, and by (5.4) we have $NT = \prod_{\rho,\sigma} NT$. If y is a point of B and if we choose $T = \delta Y$, NT is a harmonic form whose singularity at y is the same as that of $\delta_y g(x,y)$ where g is the kernel of a Green's operator.

BIBLIOGRAPHY

[1] BERGMAN, S., a) "Sur les fonctions orthogonales de plusieurs variables complexes," Mémorial des Sciences Mathématiques, vol. 106, Paris, 1947.
b) "The kernel function and conformal mapping," Mathematical Surveys, No. V., New York, 1950.

[2] BOCHNER, S., a) "Remarks on the theorem of Green," Duke Math. Journal, 3 (1937), 334-338.
b) "Analytic and meromorphic continuation by means of Green's formula," Annals of Math., 44 (1943), 652-673.
c) "On compact complex manifolds," Journ. Indian Math. Soc., 11 (1947), Nos. 1 and 2.
d) "Vector fields on complex and real manifolds," Annals of Math., 52 (1950), 642-649.
e) "Tensor fields with finite bases," Annals of Math., 53 (1951), 400-411.

[3] CALABI, E., SPENCER, D. C., "Completely integrable almost complex manifolds," Annals of Math., (to appear).

[4] DUFF, G. F. D., "Boundary value problems associated with the tensor Laplace equation," Canadian Journal of Math. (to appear).

[5] DUFF, G. F. D., SPENCER, D. C., a) "Harmonic tensors on manifolds with boundary," Proc. Nat. Acad. Sci., U.S.A., 37 (1951), 614-619.
b) "Harmonic tensors on Riemannian manifolds with boundary," Annals of Math., 56 (1952), 128-156.

[6] GARABEDIAN, P. R., "A new formalism for functions of several complex variables," Journal d'Analyse Mathématique, 1 (1951), 59-80.

[7] GARABEDIAN, P. R., SPENCER, D. C., a) "Complex boundary value problems," Technical Report No. 16, Stanford University, California, April 27, 1951.
b) "A complex tensor calculus for Kähler manifolds," Technical Report No. 17, Stanford University, California, May 21, 1951.

[8] HODGE, W. V. D., a) "Harmonic integrals," Cambridge University Press, 1941.
b) "Differential forms on a Kähler manifold," Proc. Camb. Phil. Soc., 47 (1951), 504-517.

[9] KODAIRA, K., a) "Harmonic fields in Riemannian manifolds," Annals of Math., 50 (1949), 587-665.
b) "The theorem of Riemann-Roch on compact analytic surfaces," American Journal of Math., 73 (1951) 813-875.

[10] de RHAM, G., KODAIRA, K., "Harmonic integrals," (Mimeographed Notes), Institute for Advanced Study, Princeton, 1950.

[11] SCHWARTZ, L., "Théorie des distributions," vol. I, II, Paris, 1950-51.

[12] SPENCER, D. C., "Cauchy's formula on Kähler manifolds," Proc. Nat. Acad. Sci., U.S.A., 38 (1952), 76-80.

[13] WEIL, A., "Sur la théorie des formes differentielles attachées à une variété analytique complexe," Commentarii Math. Helvetici, 20 (1947), 110-116.

[14] WEYL, H., "Die Idee der Riemannschen Fläche," Berlin, 1913.

MULTIVALUED SOLUTIONS OF LINEAR PARTIAL DIFFERENTIAL EQUATIONS

Stefan Bergman

§1. INTRODUCTION

In analogy to the study of functions satisfying ordinary differential equations in the theory of analytic functions of one complex variable, the investigation of solutions of partial differential equations seems to be a very promising chapter in the theory of functions of several complex variables.

These investigations can proceed in two directions: one can consider the solutions either in the space of n complex variables z_k, or in certain submanifolds, e.g., in the real subspace, $\text{Im } z_k = 0$, $k = 1,2,\ldots,n$.

In studying special classes of functions of two complex variables we can use general methods (e.g., the method of the kernel function, see [4, 12, 16, 17]). It seems, however, of considerable interest in addition to these general methods, also to employ more specific ones, in particular, the integral operator method. Since in the present paper we intend to explain the basic idea of this procedure and to indicate some of its characteristic features, we shall limit ourselves to a discussion of its application to the case of functions of two complex variables and we shall formulate the results for the behavior of solutions only in the real plane.

One can easily generalize most of the results by considering the solutions either in other non-analytic manifolds or in the whole space of two complex variables. We shall also limit ourselves to questions connected with the theory of multivalued solutions, referring for application of the method in other directions to previous publications.

As has been shown in [3, 13], the method of mapping the algebra of analytic functions of a complex variable into the linear space of harmonic functions of two variables can be extended to the theory of general linear partial differential equations in two (and more) variables. In the case of equations of two variables, the mapping is realized by

$$\psi(z,z^*) = \text{Re}[P(f)], \quad \text{Re} = \text{Real part}, \tag{1.1}$$

where the integral operator P is given by

$$P\ (f) = \int_{-1}^{+1} E(z,z^*,t) \quad f(\tfrac{1}{2}\, z(1-t^2))\ dt/(1-t^2)^{1/2}. \tag{1.2}$$

As mentioned before, we shall consider the solutions in the real plane; its cartesian coordinates will be denoted by x_1, x_2. Further,

$$z = x_1 + ix_2, \quad z^* = x_1 - ix_2 .$$

REMARK. When functions are considered in the whole space of two complex variables z_1, z_2; $z = z_1+iz_2$, $z^* = z_1-iz_2$ are two <u>independent</u> complex variables. $E(z,z^*,t)$ (the so-called generating function of the operator) is a <u>fixed</u> function (which depends only upon the equation), while f is an arbitrary function of a complex variable z which is regular at the origin. (Note that if $y_1=y_2=0$, $z^*=\bar{z}$ is the conjugate of z.)

In some instances it is more convenient to write the operator in the form

$$\psi(z,z^*) = \operatorname{Re}[p(g)] \tag{1.3}$$

where

$$f(\tfrac{z}{2}) = -\frac{1}{2\pi} \int_{-1}^{1} g(z(1-t^2)) \ t^{-2}\, dt + \text{const.}$$

and conversely,

$$g(z) = \int_{-1}^{1} f(\tfrac{1}{2} z(1-t^2))\, dt/(1-t^2) \tag{1.4}$$

See [7 p. 135].

g is often denoted as the <u>associate</u> of u with respect to the operator p. For every differential equation there exist infinitely many operators (1.2), i.e., infinitely many E which yield solutions of the same equation. Some of them have very useful properties. In particular, the so-called operator of the first kind (which corresponds to a certain choice of E) is of great interest, since it preserves many properties of the function to which it is applied. (See §2 and [9, p. 299], see also [18, 19, 20, 21, 22].)

By (1.1) (or (1.2)) the associate of a given function is only defined locally (in the neighborhood of the origin). By analytic continuation this connection is defined in the large in the intersection of the regularity domains of u and f. (It can, of course, happen that the continuation may lead to multivalued functions as in the case of $g = \log(z-\alpha)$ and $\psi = \log|z-\alpha|$).

For integral operators of the first kind (as well as for some other operators) one can obtain a number of different representations, see

(2.3a)-(2.3c), which fact is of importance for the theory, since different forms of operators exhibit various properties preserved when passing from the associate to the solution. See also Remark, p. 230.

In particular, the integral operators of the first kind can be used for the study of multivalued solutions. If g in (1.3) is a multivalued function, then $\psi(z,z^*)$ will be (in general) also multivalued. In the case of special integral operators, (1.3) can be written also in the forms (2.3b) (or (2.3a)), which representation shows how various properties of g are preserved, or slightly modified, in this transition.* Thus many chapters of the theory of functions of one complex variable can be easily utilized in the theory of multivalued solutions of differential equations. In particular, in [7], we indicate the connections between the properties of subsequences $\{a_{m0}\}$ of the development of

$$\psi(z,z^*) = \sum_{m,n=0}^{\infty} a_{mn} z^m z^{*n}$$

on one side, and those of the function ψ on the other. (These results represent a generalization of the Weierstrass-Hadamard direction). Further, exploiting the classical theory of the integrals of algebraic functions, we obtain series developments for multivalued solutions $\psi(z,z^*)$.

The case of differential equations with singular and single-valued coefficients is discussed in §3. In §4, we consider differential equations of mixed type, which we investigate using another type of integral operator (the so-called integral operator of the second kind).

In the present paper we formulate theorems for <u>real</u> variables, x_1,x_2, since in previous papers the results are in most instances discussed in the real domain. It should however be stressed that the continuation and study of functions obtained in this way in complex domains (i.e., replacing x_k by $z_k = x_k + iy_k$, $k = 1,2$) in most of the considerations in §2 and 3, does not involve any essential difficulties. In this way we arrive at results about functions of two complex variables satisfying a linear partial differential equation.

In [8] it has been shown how the method of integral operators can be generalized to the case of linear differential equations of the fourth order. The considerations of [8] can be extended more generally to the differential equations of $2n^{th}$ order, $n > 2$, so that the method discussed in Sections 2 and 3 can be extended to the case of linear partial differential equations of $2n^{th}$ order. In [10, 11, 13] is discussed how to generalize the methods of mapping solutions of certain differential

---

* Roughly speaking, due to this fact the theory of Riemann surfaces of one complex variable yields not only the theory of these surfaces for harmonic functions, but generally for solutions of equations (2.2) with entire coefficients A, B, F.

equations in three variables, namely of $\Delta\psi = 0$, $\Delta\psi + F(r^2)\psi = 0$, $r^2 = x_1^2 + x_2^2 + x_3^2$ and of harmonic vectors (i.e., solutions of curl $\vec{H} = 0$, div $\vec{H} = 0$) onto algebras. These mappings again can be realized by certain integral operators, and they permit us to study multivalued solutions in the real or in the complex domain.

It seems that the results obtained are only special cases of much more general theorems and that the method of integral operators opens among other things vast possibilities for the general study of multivalued solutions of linear partial differential equations in n variables, $n > 2$.

§2. INTEGRAL OPERATORS OF THE FIRST KIND IN THE CASE OF EQUATIONS OF THE FORM (2.1) WITH ENTIRE F

An integral operator transforms analytic functions of one complex variable into solutions of the differential equations

$$\frac{1}{4}\Delta\psi + F(x_1, x_2)\psi = 0, \tag{2.1}$$

$$\Delta\psi \equiv \frac{\partial^2\psi}{\partial x_1^2} + \frac{\partial^2\psi}{\partial x_2^2} \equiv 4\,\frac{\partial^2\psi}{\partial z\,\partial z^*}, \qquad z = x_1 + ix_2, \qquad z^* = x_1 - ix_2\ .$$

(In the present paper for the sake of brevity, we consider only the equation (2.1). We note that the methods and results can be easily generalized to the case of the equation

$$\Delta\psi + A\psi_{x_1} + B\psi_{x_2} + F\psi = 0 \tag{2.2}$$

where A, B, F are entire functions.) In this section we shall assume that F is an entire function. The integral operator of the first kind p_1 (which corresponds to a special choice of E in (1.1)) can be also written in the form

$$p_1(g) = g(z) + \sum_{n=1}^{\infty} \frac{\Gamma(2n+1)Q^{(n)}(z,z^*)}{2^{2n}\,\Gamma(n+1)} \int_0^z \int_0^{z_1} \cdots \int_0^{z_{n-1}} g(z_n)\,dz_n \ldots dz_1 \tag{2.3a}$$

$$= g(z) + \sum_{n=1}^{\infty} \frac{(-1)^{n-1}Q^{(n)}(z,z^*)}{2^{2n}\,B(n,n+1)} \int_0^z (z-\zeta)^{n-1}\,g(\zeta)\,d\zeta \tag{2.3b}$$

$$(2.3c) \quad = g(z) - \int_0^z \int_0^{z^*} \widetilde{F} g \, dz_1 \, dz_1^* + \int_0^z \int_0^{z^*} \widetilde{F} \int_0^{z_1} \int_0^{z_1^*} \widetilde{F} g \, dz_2 dz_2^* \, dz_1 dz_1^* + \\ + \cdots$$

$\widetilde{F} \equiv F((z + z^*)/2, \quad (z - z^*)/2i)$, Γ = the gamma function, B = the beta function. Here $Q^{(n)}$ are certain fixed functions which depend only on F and are independent of g. In the case of the integral operator of the first kind we have $Q^{(n)}(z,0) = 0$, and

$$g(z) = 2\psi(z,0) + \text{const.} \tag{2.4}$$

REMARK. The simultaneous use of all three forms ((2.3a)-(2.3c)) of the integral operator is of considerable use in many investigations. The formulas (2.3a) and (2.3b) have the advantage that the influence of the coefficient F of the equation on the property of the solution $p_1(g)$ is "separated" from the influence of the associate g. On the other hand, in the case of singular F the domain of validity of (2.3c) is, in general, larger than that of (2.3a) and (2.3b).

Among other useful properties*, the integral operator of the first kind has the following which are of considerable importance for the theory of multivalued solutions:

I. The associate g is regular in every simply-connected domain (which includes the origin) in which the solution ψ is regular.

If g has a pole at P, then $p_1(g)$ is infinite of the same order at P but has there (in general) a branch point of infinite order. (These singularities are denoted as pole-like singularities). If g has at P an algebraic branch points, $p_1(g)$ has at P a branch point of the same order.

II. If a (real) solution ψ has the development

$$\psi(z,z^*) = \sum_{m,n=0}^{\infty} a_{mn} z^m z^{*n} , \tag{2.5}$$

then, in accordance with (2.4), the associate is

$$g(z) = 2 \sum_{m=0}^{\infty} a_{m0} z^m + \text{const.} \tag{2.6}$$

* See, e.g., [9], where the reader will find also bibliographical data.

I and II have many consequences of interest in the theory of multi-valued solutions of (2.1):

Theorems of Weierstrass, Hadamard, Mandelbrojt, Polya, Szego, and many others, establish various relations between the coefficients of the series development of a function g of one complex variable on one side, and properties of multivalued solutions whose function element at the origin is g, on the other. In particular, various results referring to properties of Riemann surfaces on which g, (and $\mathrm{Re}\, g$) is defined are known. Due to I, II, and the representation (2.3a) of the operator p_1, these theorems can be formulated not only as theorems about harmonic functions, but more generally as theorems about solutions of differential equation (2.1) with entire F. In many instances, properties of solutions ψ obtained in this way are either completely independent of the coefficient F of the equation, or dependent only on some properties of F. For details, see [7, pp. 141, ff., 9: pp. 321 ff.].

Using the classical results in the theory of integrals of algebraic functions we obtain interesting representations for solutions ψ whose associates of the first kind are rational or algebraic functions.

If the associate $g(z)$ of $\psi(z,z^*)$ [i.e., if $\psi(z,0)$] is a rational function then using (2.3b), $p_1(g)$ can be represented in the form

$$p_1(g) = g(z) + \sum_{n=1}^{\infty} \frac{(-1)^{n-1} Q^{(n)}(z,z^*)}{2^{2n}\, B(n,n+1)}\, g_n(z) \tag{2.7}$$

where the $g_n(z)$ are algebro-logarithmic functions for which simple explicit formulas can be obtained.

In the case of algebraic associates $g(z)$ similar representations hold. Let g be an algebraic function which is single valued on the Riemann surface R_o (of genus p) defined by the irreducible equation

$$A(z,y) = A_o(z)\, y^m + A_1(z)\, y^{n-1} + \dots + A_m(z) = 0 \tag{2.8}$$

where $A_\nu(z)$ are polynomials. We can associate with R_o certain functions, and finitely many transcendental functions $J_\alpha(z,y)$ (integrals of the first kind). See [23], pp. 516, 533.

Suppose a function $N(z)$ can be represented as a finite expression involving: 1) Theta functions $\theta(u_1,\dots,u_p)$ associated with R_o; 2) their derivatives with respect to the u_ν; 3) integrals of the first kind $J_\alpha(z,y)$, $\alpha = 1,2,\dots,p$, defined on R_o; 4) finitely many algebro-logarithmic expressions. Then N will be said to belong to the class $G(R_o)$: $N \in G(R_o)$.

Let $\psi(z,z^*)$ be a (real) solution of (2.1) whose associate (i.e., $2\psi(z,0)$ + const.) is an algebraic function defined on the Riemann surface

R_o. Then $\psi(z,z^*)$ can be represented in the form (2.7) where $g_n(z) \in G(R_o)$. See [14] §§2, 3.

Similar representations can be obtained for solutions of differential equations of higher (even order). For differential equations of fourth order of the form indicated in [8], p. 619, these results follow immediately from formulas (2.4), (2.5), and (1.5) of [8].

The pole-like singularities mentioned in I are in general, i.e., if $F \neq 0$, infinitely many-valued functions (even if considered in the real plane). However there exists another type of singularity which is single valued <u>if we restrict ourselves to the real domain</u>.

The function

$$\psi^{(L)}(x_1,x_2;x_1^0,x_2^0) = \frac{1}{2} A(\log \zeta + \log \zeta^*) + B \tag{2.9}$$

where

$$\zeta = z-z_o, \quad \zeta^* = z^*-z_o^*, \quad z = x_1+ix_2, \quad z^* = x_1-ix_2, \quad z_o = x_1^0+ix_2^0,$$
$$z_o^* = x_1^0-ix_2^0 ,$$

$$A = 1 - \int_{z_o}^{z} \int_{z_o^*}^{z^*} F\, dz_1\, dz_1^* + \int_{z_o}^{z} \int_{z_o^*}^{z^*} F\left(\int_{z_o}^{z_1^*} \int_{z_o^*}^{z_1^*} F\, dz_2\, dz_2^*\right) dz_1\, dz_1^* + \cdots$$

$$B = \int_{z_o}^{z} \int_{z_o^*}^{z^*} G\, dz_1\, dz_1^* - \cdots$$

$$G = -\frac{1}{\zeta^*}\frac{\partial A}{\partial z} - \frac{1}{\zeta}\frac{\partial A}{\partial z^*}$$

becomes logarithmically infinite at the point (x_1^0,x_2^0). Differentiating (2.9) with respect to the parameters x_1^0 and x_2^0 yields single-valued $\partial^n \psi/\partial x_k^{(o)n}$ which are infinite of integer order. In analogy to the Mittag-Leffler theorem it is possible to obtain representations for solutions ψ of (2.1) which possess singularities of form (2.9) at a given set of points having an accumulation point only at infinity.

The proof of the above statement proceeds in the same way as in the case of analytic functions of one variable. We need only show that to every solution $\psi(z,z^*)$, $z^* = \bar{z}$, regular in a (closed) circle $\left[x_1^2 + x_2^2 \leq \rho^2\right]$, there exists a finite combination

$$\sum_{\nu=1}^{n} a_\nu \psi_\nu(z,z^*)$$

of entire solutions (i.e., solutions which are regular at every finite point), and such that

$$(2.10) \qquad \left|\psi - \sum_{\nu=1}^{n} a_\nu \psi_\nu\right| \leq \varepsilon \quad \text{for } x_1^2 + x_2^2 \leq \rho^2$$

for any given ε. This result follows immediately from the theorem that every solution regular in a circle can be developed there in the series

$$\psi(z,z^*) = \sum_{\nu=1}^{\infty} b_\nu \psi_\nu(z,z^*),$$

where

$$(2.11) \qquad \psi_{2n-1}(z,z^*) = \mathrm{Re}\left[p_1(z^n)\right], \quad \psi_{2n}(z,z^*) = \mathrm{Im}\left[p_1(z^n)\right].$$

[7, p. 140].

Using this procedure we obtain for a function possessing the singularities (2.9) in the set $\{x_1^{(m)}, x_2^{(m)}\}$, $m = 1,2,3,\ldots$ a representation

$$(2.12) \quad \psi(x_1,x_2) = \sum_{m=1}^{\infty}\left[\psi^{(L)}\left(x_1,x_2,x_1^{(m)},x_2^{(m)}\right) - \sum_{\nu=1}^{N_m} b_\nu^{(m)} \psi_\nu(x_1,x_2)\right].$$

We wish also to mention that the fact that the solution and its associate of the first kind have singularities at the same points leads to interesting limit relations for certain sequences of derivatives of the solution of the differential equation.

For instance, let us assume that $\psi(z,z^*)$ is an entire function. If we shift the coordinate system, i.e., write

$$(2.13) \qquad z' = z - \alpha, \quad z^{*\prime} = z^* - \bar{\alpha},$$

then we see that the associate $\sum \beta_{n0} z'^n$ of $\psi_1(z',z^{*\prime}) = \psi(z-\alpha, z^*-\bar{\alpha})$, where

$$(2.14) \qquad \beta_{n0} = \frac{1}{n!}\left(\frac{\partial^n \psi(z,z^*)}{\partial z^n}\right)_{z=\alpha,\, z^*=\bar{\alpha}} \qquad n = 0,1,2,\ldots$$

is also an entire function, which results in certain limit relations for the sequence (2.14).

It should be finally mentioned that results similar to those discussed in this section can be obtained for harmonic function in three variables, as well as for solutions of differential equations of the form

$$(2.15) \qquad \frac{\partial^2\psi}{\partial x_1^2} + \frac{\partial^2\psi}{\partial x_2^2} + \frac{\partial^2\psi}{\partial x_3^2} + F(r^2)\psi = 0, \quad r^2 = x^2 + y^2 + z^2,$$

where $F(r^2)$ is an entire function of r^2. See [10, 13, p. 499 ff.].

§3. DIFFERENTIAL EQUATION (2.1) WITH SINGULAR, SINGLE-VALUED COEFFICIENT F

In the present and in the next section we shall investigate the solutions of differential equation (2.1) assuming that the coefficient F is singular. In §3 we discuss the case where the coefficient F is single-valued while in §4 a multivalued F will be considered in certain special instances.

In this section we assume that the solutions ψ of (2.1) are defined in a bounded, simply-connected domain D which includes the origin. Here the coefficient F is an analytic function of two variables, and we shall assume that it has at some point, say $R \neq 0$, of D a singularity; otherwise, F is regular in D. As one can see from the representation (2.3c) the integral operator of the first kind transforms analytic functions $g(z)$ which are regular in $D-R$, into solutions of (2.1) which are defined on the universal covering surface S of $D-R$. One obtains the class $\{p_1(g)\}$ of the above-mentioned solutions applying operator (2.3c) to all analytic functions of one complex variable which are regular on the universal covering surface S.

> REMARK. One can distinguish in the class $\{p_1(g)\}$ two subclasses $\{p_1(g_1)\}$ and $\{p_1(g_2)\}$ where $\{g_1\}$ are analytic functions which are regular in D and $\{g_2\}$ those which are regular in $D-R$ and single-valued in D.

As we mentioned before, the simultaneous use of all three forms (2.3a)-(2.3c) of the integral operator p_1 is useful. An advantage of (2.3c) is that if F is not an entire function (2.3a) and (2.3b) will converge only in a certain subdomain of the domain of regularity of F while (2.3c) converges in S. By certain modifications of the representation (2.3b) one can obtain representations for the operator p_1 which hold in $D-R$ (and not only in a subdomain of $D-R$). Let the transformation

$$\zeta = \zeta(z) \qquad \zeta = \xi_1 + i\,\xi_2, \qquad z = x_1 + ix_2, \tag{3.1}$$

map the universal covering surface S of $D-R$ onto a schlicht domain L, say the unit circle. The function $\tilde{\psi}(\zeta, \zeta^*) = \psi\big(z(\zeta), z^*(\zeta^*)\big)$ satisfies the differential equation

$$\frac{\partial^2 \tilde{\psi}}{\partial \zeta \partial \zeta^*} + \tilde{F}_1(\zeta, \zeta^*)\ \tilde{\psi} = 0, \tag{3.2}$$

$$\tilde{F}_1(\zeta, \zeta^*) = F\left(z(\zeta), z^*(\zeta^*)\right) \frac{dz}{d\zeta} \frac{dz^*}{d\zeta^*}$$

and $\tilde{\psi}$ and $\tilde{F}_1$ are now defined in a schlicht simply-connected open domain L. Approximating F_1 by polynomials $\tilde{F}_N(\zeta, \zeta^*)$ we obtain in certain instances for the operator of the first kind the representation

$$(3.3) \qquad p_1(g) = \lim_{N \to \infty} p_1^{(N)}(g),$$

$$p_1^{(N)} = g\left(\zeta(z)\right) + \sum_{n=1}^{\infty} \frac{(-1)Q_N^{(n)}\left(\zeta(z), \zeta^*(z^*)\right)}{2^{2n} B(n,n+1)} \int_0^{\zeta(z)} (\zeta-\tau)^{n-1} g(\tau)\, d\tau$$

which is valid in S. Here $Q_N(\zeta, \zeta^*)$ are coefficients of (2.3b) which correspond to $\tilde{F}(\zeta, \zeta^*) = \tilde{F}_N(\zeta, \zeta^*)$ in (2.1).

REMARK. In using the above procedure it is sometime advantageous to use the Mittag-Loeffler formula for the approximation of F.

Various results in the theory of ordinary differential equations are based on the fact that the general solution φ of linear differential equations of nth order can be represented as a linear combination

$$(3.4) \qquad \varphi = \sum_{\nu=1}^{n} a_\nu \varphi_\nu$$

of n independent particular solutions with coefficient a_ν. In particular, the study of singularities and Riemann surfaces of solutions of linear differential equations of Fuchs type is closely connected with the study of a group of transformations of the coefficients a_ν of (3.4) if we move around the singularity of the coefficients of the equation. It seems that the theory of integral operators opens a possibility for a generalization of some procedures arising in the Riemann-Fuchs approach to the case of partial differential equations. As has been shown, the solutions

$$(3.5) \qquad \psi_{2\nu-1} = \mathrm{Re}\left[p_1(z^\nu)\right], \quad \psi_{2\nu} = \mathrm{Im}\left[p_1(z^\nu)\right], \qquad \begin{array}{l}\mathrm{Re} = \text{Real part,} \\ \mathrm{Im} = \text{Imaginary part,}\end{array}$$

play for the equation (2.1) a role similar to that of $\mathrm{Re}\, z^\nu$, $\mathrm{Im}\, z^\nu$ in the case of the Laplace equation. In particular, it is possible to show that to every equation (2.1) with F regular in D-R there exists a ρ_0 so that every solution of (2.1) which is regular in the circle $C = [x_1^2 + x_2^2 \leq \rho_0^2]$ can be developed in C in the form

$$(3.6) \qquad \psi = \sum_{\nu=1}^{\infty} a_\nu \psi_\nu$$

the series (3.6) converging in C uniformly. Each ψ_ν is regular on the universal covering surface S of D-R. Let us denote by $\psi_\nu^{(2)}(z,z^*)$, $(z,z^*) \in C$, the functions which one obtains when starting from $\psi_\nu = \psi_\nu^{(1)}$ and moving around R once. The functions $\psi_\nu^{(2)}$ are single-valued and regular in C and they can be developed there in the form

$$\psi_\nu^{(2)} = \sum_{\mu=1}^{\infty} \left(\delta_{\nu\mu} + a_{\nu\mu}\right) \psi_\mu^{(1)}, \quad \psi_\nu^{(1)} \equiv \psi_\nu, \ \nu = 1,2,\ldots,$$
$$\delta_{\nu\mu} = 1, \quad \delta_{\nu\mu} = 0, \text{ for } \nu \neq \mu. \tag{3.7}$$

To every homotopy class Γ of curves corresponds an infinite matrix M_Γ. The M_Γ form a group. The author will study these groups in a future paper.

§4. THE INTEGRAL OPERATOR OF THE SECOND KIND AND ITS PROPERTIES

Although using the integral operator of the first kind it is possible to determine many basic properties of solutions, it is of interest to investigate also other integral operators. Some of them exhibit new connections between certain properties of solutions. Often the results obtained in this manner are to a large extent independent of the coefficient F of equation (2.1).

For E to be a generating function of an integral operator yielding solutions of a given differential equation, it is only necessary that E satisfy a certain partial differential equation and certain general conditions (see [3], §1). Thus, one has considerable freedom in choosing a generating function for a given partial differential equation (2.1), Usually in defining an integral operator we impose on E some initial value conditions. For instance, in the case of the integral operator of the first kind we require

$$E(z,0,t) = E(0,z^*,t) = 1. \tag{4.1}$$

As a consequence of this choice it follows that the associates of the first kind of $\psi(z,z^*)$ is (2.4), i.e., $2\psi(z,0)$ + const. On the other hand, one can impose conditions different from (4.1) so that the associate of $\psi(z,z^*)$ is connected with $\psi(z,z^*)$ by some other law. Often the initial conditions can be chosen in such a way that the associate is either completely independent of F or depends only on some of its properties. In particular, in the present section we consider an integral operator whose generating function E_2 satisfies

$$E_2(z,-z^*,t) = 0 \tag{4.2}$$

and a second condition referring to its derivative.

This integral operator can be used for the study of equation (2.1) with a multivalued coefficient F.

This kind of coefficient appears in the study of equations of mixed type. Reducing them to the normal form, we obtain equations of the form (2.1) with an F possessing an algebraic singularity along the curve corresponding to the transition line. See [13] and [15].

In the following, we consider a special equation of mixed type, namely,

$$M(\psi) \equiv \psi_{HH} + \ell(H)\, \psi_{x_2 x_2} = 0. \tag{4.3}$$

Here $\ell(H)$ has in the neighborhood of the origin the development[1]

$$\ell(H) = \sum_{n=1}^{\infty} (-1)^n a_n H^n, \quad a_1 > 0,\ a_2 < 0,\ a_n \text{ real for } n = 3, 4, \ldots, \tag{4.4}$$

and has the property that

$$\ell(H) > 0 \text{ for } H < 0, \quad \ell(H) < 0 \text{ for } H > 0, \ . \tag{4.5}$$

Finally $\ell(H)$ is subjected to certain conditions which will be formulated below, see p. 241.

By the transformation

$$-x_1(-H) = \int_{t=0}^{-H} [\ell(t)]^{-1/2}\, dt = \tag{4.6}$$

$$\frac{2}{3} a_1^{1/2} (-H)^{3/2} + \frac{a_2}{3a_1^{1/2}} (-H)^{5/2} + \ldots$$

the equation (4.3) is reduced to the form (see (2.9) of [13])

$$L(\psi) \equiv \psi_{x_1 x_1} + \psi_{x_2 x_2} + N(x_1)\, \psi_{x_1} = 0, \tag{4.7}$$

$$N(x_1) = \frac{1}{8} \ell^{-3/2} \ell_H .$$

REMARK. In the present section we use the form (4.7) rather than (2.1), since in previous papers the form (4.7) has been used. By the transformation $\psi = R\,\psi^*$, where $R = \exp[-\int_0^{z^*} N\, dz_1^*]$, we pass from one to another form. In accordance with (4.5) and

---

[1] We assume here that $\ell(H)$ when continued to complex values of H is an analytic function of H. This assumption can be replaced by a much weaker one. It is made here only for the sake of simplicity.

(4.6) if $H < 0$, x_1 is real and negative while if $H > 0$, $-ix_1 > 0$ (x_1 is purely imaginary). These two regions correspond to regions of the elliptic and of hyperbolic character of the equation (4.3), respectively. In the present discussion, we shall limit ourselves to the consideration of solutions for $x_1 < 0$, i.e., in the domain where the equation has elliptic character. But the procedures used can easily be extended so as to obtain various results about the solutions for $H > 0$.

The following four additional assumptions will be made about the coefficient N of (4.7). (They imply conditions on $\mathcal{L}$, which were mentioned before. See [15], §3.)

1. The function N possesses an expansion of the form

$$N(x_1) = (-x_1)^{-1}\left[-\frac{1}{12} + \sum_{\nu=1}^{\infty} \beta_\nu (-x_1)^{2\nu/3}\right], \quad \beta_1 > 0, \tag{4.8}$$

which is valid for $-x_1^{(o)} < x_1 < 0$, $x_1^{(o)} > 0$.

2. $N(x_1)$ is an analytic function for $-\infty < x_1 < 0$ and is real for $x_1 < 0$.

3. The expression

$$\exp\left[-\int_{-\infty}^{x_1} 2N(t)dt\right] = h(x_1) = S_o\,(-x_1)^{-1/6}\left[1+S_1(-x_1)^{2/3}+\ldots\right] \tag{4.9}$$

exists for all $x_1 < 0$.

4. It may now be shown that

$$\lim_{x_1 \to 0^-} (-x_1)^{1/6}\, h(x_1)$$

exists. We denote this limit by S_o and require

$$S_o > 0. \tag{4.10}$$

If we choose

$$\beta_\nu = 0 \qquad \nu = 1,2,\ldots \tag{4.11}$$

so that we have for $N(x_1)$

$$N(x_1) = (12x_1)^{-1} \equiv N^{\dagger}(x_1), \tag{4.8a}$$

the so-called generating function of the second kind, $E_2 \equiv E_2^{\dagger}(x_1,x_2,t)$ becomes

$$(4.12)\quad E_2^{\dagger}(x_1,x_2,t) = AS_o(-2x_1)^{-1/6}\mathfrak{F}\Big[1/6,5/6,1/2;\big(-t^2(x_1+ix_2)\big)/(-2x_1)\Big] +$$

$$+ BS_o(-2x_1)^{2/3}\mathfrak{F}\Big[2/3,4/3,3/2;\big(-t^2(x_1+ix_2)\big)/(-2x_1)\Big],$$

where $\mathfrak{F}$ is the hypergeometric function, A and B are conveniently chosen constants. Consequently in the special case where $N = N^{\dagger}$ the solutions of (4.7) can be represented in the form

$$(4.13)\qquad \psi(z,z^*) = \mathrm{Re}\left[P_2^{\dagger}(f)\right]$$

$$(4.13a)\qquad P_2^{\dagger}(f) = \int_{t=-1}^{1} E_2^{\dagger}(x_1,x_2;t_1)\, f\left(\tfrac{1}{2}z(1-t^2)\right) dt/(1-t^2)^{1/2}\ .$$

Using the fact that in this special case $E_2^{\dagger}$ is a hypergeometric function, and that various properties of the hypergeometric function imply corresponding properties of the solutions of

$$(4.7a)\qquad \psi_{x_1x_1} + \psi_{x_2x_2} + 4N^{\dagger}(x_1)\psi_{x_1} = 0,$$

one can derive various results referring to functions satisfying (4.7a).

It is again of considerable interest that some of these properties are valid also in the general case, i.e., in the case where N satisfies the assumptions formulated in the present section (but is not necessarily $N^{\dagger}$).

An example of a possible generalization of this type is discussed in the present paper. The functions $\mathfrak{F}$ appearing in (4.12) can be developed in a series of positive powers of $[2x_1/\big(t^2(x_1 + ix_2)\big)]$. Interchanging then the order of summation and integration in (4.13a) we obtain a formula having a structure similar to that of (2.3b). This formula holds in $[2|x_1| < (x_1^2 + x_2^2)^{1/2},\ x_1 < 0]$ and yields a representation for multivalued solutions of (4.7a) which can be used for their study. As shown in [13, 15], a representation for $E_2(x_1,x_2,t)$ analogous to that described in the special case $N = N^{\dagger}$ holds in the general case.

Let $\mathfrak{C}$ be a simple curve which connects the points -1 and $+1$ of the complex t plane and lies completely in $1 \leq |t| \leq A < \infty$, A sufficiently large. For each point ζ, we denote by $C(\zeta,\mathfrak{C})$ the curve traced out by the values $(1/2)\zeta(1-t^2)$ as t traverses $\mathfrak{C}$ from -1 to $+1$. Furthermore, we denote by $T_{\mathfrak{C}}(B)$ the domain $T_{\mathfrak{C}}(B) = \sum_{\zeta\in B} C(\zeta;\mathfrak{C})$. Then we have the following result:

Let $\chi_1(x_2)$ and $\chi_2(x_2)$ be two functions of x_2 which possess representations of the form

$$\chi_j(x_2) = \sum_{\nu=0}^{\infty} a_\nu^{(j)} x_2^\nu; \quad 0 \le x_2 \le x_2^{(0)}, x_2^{(0)} > 0, \quad (j = 1,2) \quad (4.14)$$

and let

$$f(z) = \sum_{\nu=0}^{\infty} c_\nu z^{\nu+(\frac{1}{6})} =$$

$$- \left[\frac{(-2iz)^{\frac{1}{6}}}{3^{1/2} S_0^2 \operatorname{Im}[A_2 \overline{A}_1]} \right] \left[- \overline{d}_o \int_{C_2} t^{-1/3} \chi_1(\sigma) dt + \right.$$

$$\left. + \sum_{k=1}^{2} (-1)^k \overline{d}_k \int_{C_2} t^{-5/3} \chi_k(\sigma) \, dt \right] \quad (4.15)$$

$$\sigma = - 2iz \, (1-t^2)$$

$$d_o = -(\tfrac{2}{3}) \, i^{3/2} S_o A_2, \quad d_1 = -(2^{5/3}/3) \, i^{1/6} S_o S_1 A_1, \quad d_2 = -i^{1/6} S_o A_1,$$

where S_o is given by 4. p. 241, $S_1 = 3 \beta_1$ and A_1, A_2 are conveniently chosen constants. If a domain B situated in

$$[3^{1/2} |x_1| \le x_2, \; x_2 \ge 0, \; x_1 \le 0]$$

contains the segment $0 \le x_2 \le x_2^{(0)}$, and if $f(z)$ is regular in $T_z(B)$, then

$$\psi(x_1,x_2) = \operatorname{Im} \left(\sum_{n=0}^{\infty} \left[\frac{A_1 q^{(n,k)}(x_1)}{z^{n+(1/6)}} \int_{t=-1}^{1} \frac{f(\frac{1}{2} z(1-t^2))dt}{(-t)^{n+1/6}(1-t^2)^{1/2}} + \right. \right. \quad (4.16)$$

$$\left. \left. + \frac{A_2 q^{(n,k)}(x_1)}{z^{n+(1/6)}} \int_{t=-1}^{1} \frac{f((1/2)z(1-t^2))(1-t^2)^{1/6} dt}{(-t^2)^{n+5/6}} \right] \right), \quad z = x_1 + ix_2$$

is a solution of the differential equation (4.7) which is regular in B and is such that

$$\lim_{x_1 \to 0^-} \psi(x_1,x_2) = \chi_1(x_2) \quad (4.17)$$

$$\lim_{x_1 \to 0^-} (-x_1)^{1/3} \left(\frac{\partial \psi(x_1,x_2)}{\partial x_1} \right) = \chi_2(x_2) \quad (4.18)$$

The functions $q^{(n,k)}(x_1)$ are certain fixed functions which are defined for $x_1 < 0$ and depend on the differential equation [12a, 15].

This representation permits us to study the analytic continuation of $\psi(x_1,x_2)$ to a domain whose coordinates x_1, x_2 satisfy $3^{1/2} x_1 < x_2$, $x_2 > 0$, $x_1 < 0$. The problem is thereby reduced to the question of the continuation of $f(z)$.

Thus we see that the $\chi_j(x_2)$, $j = 1,2$, determine f (essentially independently of the equation) and therefore permits us to determine the location and character of singularities (which lie in $[2|x_1| < (x_1^2+x_2^2)^{1/2}$, $x_1 < 0$, $x_2 > 0]$ independently of N).

If further $f(z)$ is an algebraic function then the integrals in (4.16) can be expressed in terms of certain Θ functions, their derivatives, and finitely many integrals of the first kind. See [14], p. 7.

In this way we obtain formulas which are analogous to a certain extent to those obtained in §2. In this case the second factors $\int_{t=-1}^{1} \ldots dt$ in (4.16) depend upon the values of ψ and

$$\lim_{x_1 \to 0^-} (-x_1)^{1/3} \left(\partial \psi(x_1,x_2)/\partial x_2\right)$$

on the transition line $x_1 = 0$, and not on $\psi(z,0)$ as in the case of an integral operator of the first kind.

The author wishes to thank Mr. Robert Osserman for his help in the preparation of the present paper.

BIBLIOGRAPHY

[1] BERGMAN, S., "Zur Theorie der algebraischen Potentialfunktionen des drei-dimensionalen Raumes," Math. Ann., vol. 99, 1928, pp. 629-659, and vol. 101, 1929, pp. 534-558.

[2] __________, "Über Kurvenintegrale von Funktionen zweier komplexen Veränderlichen, die die Differentialgleichung $\Delta V + V = 0$ befriedigen," Mathematische Zeitschrift, vol. 32, 1930, pp. 386-406.

[3] __________, "Zur Theorie der Funktionen, die eine lineare partielle Differentialgleichung befriedigen," Rec. Math. (Mat. Sbornik) N. S. vol. 2, 1937, pp. 1169-1198.

[4] __________, "Sur les fonctions orthogonales de plusieur variables complexes avec les applications à la théorie des fonctions analytiques," Interscience Publishers, 1941, and Mémorial des Sciences Mathématiques, vol. 106, Paris, 1947.

[5] __________, "The hodograph method in the theory of compressible fluids," Supplement to Fluid Dynamics by von Mises and Friedrichs, Brown University, 1942.

[6] __________, "A formula for the stream function of certain flows," Proc. Nat. Acad. Sci., U.S.A., vol. 29, 1943, pp. 276-281.

[7] __________, "Linear operators in the theory of partial differential equations," Trans. Amer. Math. Soc., vol. 53, 1943, pp. 130-155.

[8] __________, "Solution of linear partial differential equations of the fourth order," Duke Math. J., vol. 11, 1944, pp. 617-649.

[9] ___________, "Certain classes of analytic functions of two real variables and their properties," Trans. Amer. Math. Soc., vol. 57, 1945, pp. 299-331.

[10] ___________, "A class of harmonic functions in three variables and their properties," Trans. Amer. Math. Soc., vol. 59, 1946, pp. 216-247.

[11] ___________, "Classes of solutions of linear partial differential equations in three variables," Duke Math. J., vol. 13, 1946, pp. 419-458.

[12] ___________, "Sur la fonction-noyau d'un domaine et ses applications dans la theorie des transformations pseudo-conformes," Memorial des Sciences Mathematiques, vol. 108, Paris, 1948).

[12a] ___________, "Two-dimensional transonic flow patterns," Amer. J. of Math., vol. 70, 1948, pp. 856-891.

[13] ___________, "On solutions with algebraic character of linear partial differential equations," Transactions of the Amer. Math. Soc., vol. 68, 1950, pp. 461-507.

[14] ___________, "The coefficient problem in the theory of linear partial differential equations," Transactions of the Amer. Math. Soc., vol. 73, 1952, pp. 1-34.

[15] ___________, "On solutions of linear partial differential equations of mixed type," Amer. Journal of Mathematics, vol. 74, 1952, pp. 444-474.

[16] BERGMAN, S., SCHIFFER, M., "A representation of Green's and Neumann's functions in the theory of partial differential equations of second order," Duke Math. J., vol. 14, 1947, pp. 609-638.

[17] ___________, "Kernel functions in the theory of partial differential equations of elliptic type," Duke Math. J., vol. 15, 1948, pp. 535-566.

[18] EICHLER, M. M. E., "On the differential equation $u_{xx} + u_{yy} + N(x)u = 0$," Trans. Amer. Math. Soc., vol. 65, 1949, pp. 259-278.

[19] INGERSOLL, B. M., "The regularity domains of solutions of linear partial differential equations in terms of the series development of the solution," Duke Math. J., vol. 15, 1948, pp. 1045-1056.

[20] MARDEN, M., "A recurrence formula for the solutions of certain differential equations given by their series development," Bull. Amer. Math. Soc., vol. 50, 1944, pp. 208-217.

[21] MITCHELL, J., "Some properties of solutions of partial differential equations given by their series development," Duke Math. J., vol. 13, 1946, pp. 87-104.

[22] NIELSEN, K. L., "Some properties of functions satisfying partial differential equations of elliptic type," Duke Math. J., vol. 11, 1944, pp. 121-137.

[23] WEIERSTRASS, K., "Vorlesungen über die Theorie der Abelschen Transcendenten, Gesammelte Werke," Berlin, 1902.

This work was done under a contract with the Office of Naval Research.

THE THEOREM OF RIEMANN-ROCH FOR ADJOINT SYSTEMS ON KÄHLERIAN VARIETIES

Kunihiko Kodaira

§1. INTRODUCTION

Let $\mathfrak{M}_n$ be a n-dimensional compact Kählerian variety i.e. a compact complex analytic variety of complex dimension n with a Kählerian metric $ds^2 = 2\,\Sigma\, g_{\alpha\bar{\beta}}\,(dz^{\alpha}\, d\bar{z}^{\beta})$, where $(z^1, z^2, \ldots, z^n)$ denotes the system of local analytic coordinates on $\mathfrak{M}_n$. A <u>divisor</u> on $\mathfrak{M}_n$ is, a $(2n - 2)$-cycle $D = \Sigma\, m_\nu S_\nu$ with integral coefficients m_ν composed of a finite number of irreducible analytic subvarieties S_ν of $\mathfrak{M}_n$ of complex dimension $n - 1$. Each S_ν associated with $m_\nu \neq 0$ is called a <u>component</u> of D. The set of all divisors on $\mathfrak{M}_n$ constitute an additive group. Every meromorphic function F on $\mathfrak{M}_n$ which is not identically zero determines its divisor in a well known manner. The divisor of F will be denoted by (F). We say that a divisor D' is <u>linearly equivalent</u> to D and write $D' \approx D$ if there exists on $\mathfrak{M}_n$ a meromorphic function F with $(F) = D' - D$. Obviously the linear equivalence thus defined is an equivalence relation and therefore the set of all divisors can be decomposed into mutually disjoint equivalence classes. Each equivalence class is called a <u>divisor class.</u> We say that $D = \Sigma\, m_\nu S_\nu$ is <u>effective</u> and write $D \geqq 0$ if all coefficients m_ν are $\geqq 0$. A meromorphic function F on $\mathfrak{M}_n$ is called a multiple of D if $(F) - D \geqq 0$ or F is identically zero. We denote by $\mathfrak{F}(D)$ the set of all meromorphic functions on $\mathfrak{M}_n$ which are multiples of $-D$. $\mathfrak{F}(D)$ is a finite dimensional linear space over the field $\mathfrak{K}$ of all complex numbers. For any divisor D, we denote by $|D|$ the set of all <u>effective</u> divisors which are linearly equivalent to D. Such a set $|D|$ is called a <u>complete linear system</u>. Let $\{F_0, F_1, \ldots, F_h\}$ be a base of the linear space $\mathfrak{F}(D)$. Again, denote by $\mathfrak{S}_h$ an h-dimensional projective space (over the field $\mathfrak{K}$) and by $(\lambda_0, \lambda_1, \ldots, \lambda_h)$ the homogeneous coordinates of a point $\lambda \in \mathfrak{S}_h$. Then, setting $D_\lambda = (F_\lambda) + D$, $F_\lambda = \lambda_0 F_0 + \lambda_1 F_1 + \ldots + \lambda_h F_h$, we infer that the complete linear system $|D|$ consists of all divisors D_λ: $|D| = \{D_\lambda \mid \lambda \in \mathfrak{S}_h\}$. Clearly the correspondence between D_λ and λ is one-to-one; thus D_λ depends on h independent parameters. This number h is called the <u>dimension</u> of the complete linear system $|D|$ and is denoted by $\dim |D|$. Obviously $\dim |D| = \dim \mathfrak{F}(D) - 1$. Every meromorphic n-ple differential W on $\mathfrak{M}_n$ determines its divisor (W) in an obvious manner, provided that W does not vanish identically. The divisor $K = (W)$ of an arbitrary n-ple differential W not vanishing identically is called a <u>canonical divisor</u> on $\mathfrak{M}_n$. The set $|K|$ of all canonical divisors K on $\mathfrak{M}_n$ constitutes a divisor class. The corresponding complete linear system $|K|$ is called the <u>canonical system</u> on $\mathfrak{M}_n$. For an arbitrary divisor D on $\mathfrak{M}_n$, the complete linear system $|K + D|$ is called the <u>adjoint system</u> of D, provided that at least one canonical divisor K exists on $\mathfrak{M}_n$. In case there exists on $\mathfrak{M}_n$ no canonical divisor, we define the adjoint system $|K + D|$ to be an empty set.

The problem concerning the theorem of Riemann-Roch[1] consists in expressing the dimension of the complete linear system $|D|$ in terms of numerical characteristics of D and $\mathfrak{M}_n$. The main purpose of the present paper is to give an answer to this problem in the case where $|D|$ is the adjoint system $|K + S|$ of an arbitrary surface S with only ordinary singularities on a 3-dimensional compact Kählerian variety[1a] $\mathfrak{M}_3$.

§2. AN EXISTENCE THEOREM

Let $\{\Phi\}$ be the linear space consisting of all (2n-p)-forms Φ of class C^∞ on $\mathfrak{M}_n$. Then, a <u>p-current</u>[2] on $\mathfrak{M}_n$ is, by definition, a linear functional $T[\Phi]$ defined on Φ which is continuous in the following sense: For an arbitrary sequence $\Phi^{(1)}, \Phi^{(2)}, \ldots, \Phi^{(m)}, \ldots$ of (2n-p)-forms $\Phi^{(m)} \in \{\Phi\}$ such that all $\Phi^{(m)}$ vanish outside one and the same subset of $\mathfrak{M}_n$ covered by a single system of real local coordinates $(x^1, x^2, \ldots, x^{2n})$, we have $T[\Phi^{(m)}] \to 0$ $(m \to \infty)$ if each partial derivative

$$\partial^s \Phi^{(m)}_{jk\ldots\ell} \, \partial x^p \partial x^q \ldots \partial x^r$$

of each coefficient $\Phi^{(m)}_{jk..\ell}$ of $\Phi^{(m)}$ converges uniformly to zero for $m \to \infty$. For example, we can consider an arbitrary divisor $D = \Sigma\, m_\nu S_\nu$ as a 2-current by identifying D with the linear functional $D[\Phi] = \Sigma\, m_\nu \int_{S_\nu} \Phi$. The exterior derivative dT and the adjoint *T of a p-current T are defined by $dT[\Phi] = (-1)^{p+1} T[d\Phi]$, $(*T)[\Phi] = (-1)^p T[*\Phi]$, respectively. Furthermore δT and ΔT are defined by $\delta T = -*d*T$, $\Delta T = (d\delta + \delta d)T$.

---

[1] The theorem of Riemann-Roch on algebraic surfaces is well known. See Zariski [16], pp. 63-76. Recently the author has proved the corresponding theorem for complete linear systems $|D|$ on an arbitrary Kählerian surface $\mathfrak{M}_2$ in the case where D has no multiple components. See Kodaira [3], p. 852.

[1a] This paper has been presented to the conference as a preliminary report of a more detailed exposition: Kodaira [4].

[2] de Rham and Kodaira [9], Chap. II. For the theory of currents on Kählerian varieties, see Spencer [11]. Cf. also Kodaira [3], §1, where the reader will find a summary of the results in the theory of harmonic integrals which are necessary for the present paper.

We say that a p-current T vanishes at a point $\mathfrak{p}$ if there exists a neighborhood $U(\mathfrak{p})$ of $\mathfrak{p}$ such that $T[\Phi] = 0$ for all Φ vanishing outside $U(\mathfrak{p})$. By the carrier of T will be meant the set consisting of all points $\mathfrak{p}$ on $\mathfrak{M}_n$ such that T does not vanish at $\mathfrak{p}$. For example, the carrier of a divisor $D = \Sigma\, m_\nu S_\nu$ is the sum $\Sigma_{m_\nu \neq 0} S_\nu$ of all components S_ν of D. Every p-current T can be written in the form

$$T = \sum_{r+s=p} \frac{1}{r!s!} \Sigma\, T_{\alpha_1 \ldots \alpha_r \bar{\beta}_1 \ldots \bar{\beta}_s} dz^{\alpha_1} \ldots dz^{\alpha_r} d\bar{z}^{\beta_1} \ldots d\bar{z}^{\beta_s},$$

where the coefficients $T_{\alpha_1 \ldots \alpha_r \bar{\beta}_1 \ldots \bar{\beta}_s}$ are distributions on the space of local analytic coordinates $z^1, z^2, \ldots z^n$. By using this expression, we define the operators $\Pi_{r,s}$ and Λ as follows:

$$\Pi_{r,s} T = \frac{1}{r!s!} \Sigma\, T_{\alpha_1 \ldots \alpha_r \bar{\beta}_1 \ldots \bar{\beta}_s} dz^{\alpha_1} \ldots dz^{\alpha_r} d\bar{z}^{\beta_1} \ldots d\bar{z}^{\beta_s},$$

$$\Lambda T = \sum_{r+s=p-2} \frac{1}{r!s!} \Sigma\, (-1)^r\, i g^{\alpha\bar{\beta}}\, T_{\alpha\alpha_1 \ldots \alpha_r \bar{\beta}\bar{\beta}_1 \ldots \bar{\beta}_s} dz^{\alpha_1} \ldots dz^{\alpha_r} d\bar{z}^{\beta_1} \ldots d\bar{z}^{\beta_s}.$$

A p-current T is said to be of the type (r,s) if $T = \Pi_{r,s} T$. Let $\{e_1, e_2, \ldots, e_b\}$ be a base of the linear space consisting of all harmonic p-forms of the first kind on $\mathfrak{M}_n$ which is normalized in the sense that $\int_{\mathfrak{M}_n} e_j \cdot *\bar{e}_k = \delta_{jk}$. With the help of this base, we associate with any p-current T the harmonic p-form $HT = \Sigma_j\, T[*\bar{e}_j] e_j$. HT is called the harmonic part of T. As was shown by de Rham[3], the Laplace equation $\Delta X = T - HT$ has one and only one solution X satisfying $HX = 0$ (where the solution X is also a p-current). Denoting this unique solution X by GT, we introduce Green's operator[4] G. Then we have $\Delta GT = G\Delta T = T - HT$ and $GHT = HGT = 0$ for an arbitrary p-current T. Incidentally, a p-form $\Phi = (1/p!)\Sigma\, \Phi_{\alpha_1\alpha_2 \ldots \alpha_p} dz^{\alpha_1} dz^{\alpha_2} \ldots dz^{\alpha_p}$ of type (p,o) is said to be holomorphic [or meromorphic] in a domain $\mathfrak{D} \subseteq \mathfrak{M}_n$, if, at each point $\mathfrak{p}$ in $\mathfrak{D}$, the coefficients $\Phi_{\alpha_1 \ldots \alpha_p}$ are holomorphic [or meromorphic] functions of the local coordinates $z^1, z^2, \ldots, z^n$. By a p-ple differential we shall mean a meromorphic p-form Φ defined in the whole of $\mathfrak{M}_n$ satisfying $d\Phi = 0$ except for the singular points of Φ. A p-ple differential Φ is said to be of the first kind if Φ is holomorphic everywhere in $\mathfrak{M}_n$.

[3] de Rham and Kodaira [9], p. 65, Theorem 1.

[4] de Rham and Kodaira [9], pp. 63-66.

THEOREM 1. (Existence Theorem)[5]. Let T be a current of type $(p,1)$ on $\mathfrak{M}_n$, p being a positive integer $\leqq n$. If T satisfies $dT = 0$ and $HT = 0$, then $\Theta = (d\Lambda + i\delta)GT$ is a current of type $(p,0)$ on $\mathfrak{M}_n$ and satisfies $d\Theta = iT$; moreover Θ is a holomorphic p-form outside the carrier of T.

This Theorem 1 can be regarded as a generalization of the classical existence theorem on compact Riemann surfaces[6]. We remark here that the singularities of the "analytic" differential Θ is determined by the relation $d\Theta = iT$.

As an example of applications of the above existence theorem, we shall prove the existence of a multiplicative meromorphic function with given divisor[7]. Let D be a given divisor on $\mathfrak{M}_n$ which is homologous to zero. Then, considering D as a 2-current, we have $dD = HD = 0$. Moreover D is a 2-current of type $(1,1)$. Hence, by the above Theorem 1, $\Theta = (d\Lambda + i\delta)GD$ is an "analytic" simple differential which is regular outside the carrier of D. Furthermore we infer from the relation $d\Theta = iD$ that each component S_ν of the divisor $D = \Sigma m_\nu S_\nu$ is a logarithmic polar variety of the simple Picard integral $\int 2\pi\Theta$ with the residue m_ν. Consequently $\exp(\int 2\pi\Theta)$ is a multiplicative meromorphic function with the divisor D.

§3. ADJOINT SYSTEMS ON n-DIMENSIONAL VARIETIES

For an arbitrary divisor D on $\mathfrak{M}_n$, we denote by $\mathfrak{W}(D)$ the linear space consisting of all n-ple differentials which are multiples of $-D$. Then, assuming that there exists on $\mathfrak{M}_n$ at least one n-ple differential $W_0 (\neq 0)$ and setting $K = (W_0)$, we infer readily that the correspondence $F \to W = FW_0$ between $F \in \mathfrak{F}(K + D)$ and $W \in \mathfrak{W}(D)$ is one-to-one. Hence we get

$$\dim |K + D| = \dim \mathfrak{W}(D) - 1. \tag{3.1}$$

This relation (3.1) is valid also in the case where $\mathfrak{M}_n$ has no n-ple differentials other than 0, since, in this case, the dimension of the empty set $|K + D|$ is equal to -1. Incidentally we denote for an arbitrary compact Kählerian variety $\mathfrak{M}_n$ the number of linearly independent λ-ple differentials of the first kind on $\mathfrak{M}_n$ by $r_\lambda(\mathfrak{M}_n)$ and set

---

[5] Kodaira [3], Theorem 1.4,

[6] Weyl [15].

[7] Weil [14], de Rham and Kodaira [9], Part II.

$$a(\mathfrak{M}_n) = r_n(\mathfrak{M}_n) - r_{n-1}(\mathfrak{M}_n) + r_{n-2}(\mathfrak{M}_n) - + \dots + (-1)^{n-1} r_1(\mathfrak{M}_n). \tag{3.2}$$

Now, let S be an irreducible non-singular subvariety of $\mathfrak{M}_n$ of complex dimension $n - 1$. In order to deduce the theorem of Riemann-Roch for the adjoint system $|K + S|$ of S, it is sufficient to compute the dimension of the linear space $\mathfrak{W}(S)$, as (3.1) shows. Denote by $\mathfrak{w}$ the linear space consisting of all $(n-1)$-ple differentials w of the first kind on S and associate with each $w \in \mathfrak{w}$ the $(n+1)$-current $T(w)$ of type $(n,1)$ on $\mathfrak{M}_n$ defined by $T(w)[\Phi] = \int_S w \cdot \Phi_S$, where Φ is a "variable" $(n-1)$-form of class C^∞ on $\mathfrak{M}_n$ and Φ_S denotes the $(n-1)$-form induced by Φ on S. Then, as one readily infers, $T(w)$ satisfies $dT(w) = 0$. Each n-ple differential $W \in \mathfrak{W}(S)$ can be considered as an n-current of type $(n,0)$ on $\mathfrak{M}_n$, since the integral $W[\Psi] = \int_{\mathfrak{M}_n} W \cdot \Psi$ converges absolutely for an arbitrary n-form Ψ of class C^∞ on $\mathfrak{M}_n$. Now, it can be shown[8] that the exterior derivative dW of this n-current W is represented in the form $dW = 2\pi i T(w)$, where $w \in \mathfrak{w}$. The $(n-1)$-ple differential w determined uniquely by the relation $dW = 2\pi i T(w)$ is called the <u>residue</u> of W on its polar surface S. The residue w of W will be denoted by $\mathfrak{R}(W)$. Obviously the relation $dW = 2\pi i T(w)$ implies that $HT(w) = 0$. Moreover we can prove, with the help of Theorem 1, that, if conversely, for given $w \in \mathfrak{w}$, the $(n+1)$-current $T(w)$ satisfies $HT(w) = 0$, the n-current $W = 2\pi(d\Lambda + i\delta)GT(w)$ is an n-ple differential belonging to $\mathfrak{W}(S)$ and satisfies $dW = 2\pi i T(w)$. Thus, for given $w \in \mathfrak{w}$, there exists $W \in \mathfrak{W}(S)$ with $\mathfrak{R}(W) = w$ if and only if $T(w)$ satisfies $HT(w) = 0$. It is easy to see that the condition $HT(w) = 0$ is equivalent to the following one:

$$\int_S w \cdot \overline{B}_S = 0 \tag{3.3}$$

for all $(n-1)$-ple differentials B of the first kind on $\mathfrak{M}_n$. Denoting by $\mathfrak{w}_0$ the subspace of $\mathfrak{w}$ consisting of all $w \in \mathfrak{w}$ satisfying (3.2), we infer therefore that $W \rightarrow w = \mathfrak{R}(W)$ is a linear mapping which maps $\mathfrak{W}(S)$ <u>onto</u> $\mathfrak{w}_0$. The kernel of this mapping $W \rightarrow w = \mathfrak{R}(W)$ is the space $\mathfrak{W}(0)$ consisting of all n-ple differentials of the first kind on $\mathfrak{M}_n$. Hence we obtain

$$\dim \mathfrak{W}(S) = r_n(\mathfrak{M}_n) + \dim \mathfrak{w}_0. \tag{3.4}$$

Denote by ℓ the number of linearly independent $(n-1)$-ple differentials B of the first kind on $\mathfrak{M}_n$ such that $B_S = 0$. Then the number of linearly independent conditions involved in (3.3) is equal to $r_{n-1}(\mathfrak{M}_n) - \ell$ and

[8] Kodaira [5], §3.

consequently $\dim \mathfrak{w}_0 = r_{n-1}(S) - r_{n-1}(\mathfrak{M}_n) + \ell$. Inserting this into (3.4) and combining it with (3.1), we get the following

THEOREM 2[9]. Let S be an irreducible non-singular subvariety of $\mathfrak{M}_n$ of dimension $n-1$. Then the dimension of the adjoint system $|K + S|$ of S is given by

$$\dim |K + S| = r_n(\mathfrak{M}_n) - r_{n-1}(\mathfrak{M}_n) + r_{n-1}(S) + \ell - 1, \tag{3.5}$$

where ℓ is the number of linearly independent $(n-1)$-ple differentials B of the first kind on $\mathfrak{M}_n$ which vanish on S in the sense that $B_S = 0$, B_S being the $(n-1)$-ple differential on S induced by B.

This Theorem 2 can be regarded as the theorem of Riemann-Roch for the adjoint system $|K + S|$ of S.

Now we consider the case in which $\mathfrak{M}_n$ is an irreducible non-singular algebraic variety imbedded in a projective space. Let E be a general hyperplane section[10] of $\mathfrak{M}_n$. Then, applying formula (3.5) to $|K + E|$, we get $\dim |K + E| = r_n(\mathfrak{M}_n) - r_{n-1}(\mathfrak{M}_n) + r_{n-1}(S) - 1$, since, in thi case, $\ell = 0$, while it follows from a theorem of Lefschetz[11] that $r_\lambda(E) = r_\lambda(\mathfrak{M}_n)$ for $\lambda \leqq n - 2$. Hence we obtain the following

THEOREM 3[12]. The dimension of the adjoint system $|K + E|$ of a general hyperplane section E of $\mathfrak{M}_n$ is given by

$$\dim |K + E| = a(\mathfrak{M}_n) + a(E) - 1. \tag{3.6}$$

It is not difficult to generalize the formula (3.5) to the case in which S consists of two irreducible non-singular components S_1, S_2 such that the intersection $S_1 \cap S_2$ is also an irreducible non-singular variety with the intersection-multiplicity 1. In particular we can prove the following

---

[9] Kodaira [5], Theorem 1.

[10] It is well known that a general hyperplane section is an irreducible non-singular subvariety.

[11] Lefschetz [7], pp. 88-91.

[12] Kodaira [5], Theorem 3.

THEOREM 4[13]. Let S be an irreducible non-singular subvariety of $\mathfrak{M}_n$ of dimension $n-1$ and E be a general hyperplane section of $\mathfrak{M}_n$ cutting out on S an irreducible non-singular variety $E \cap S$ with the intersection multiplicity 1. Then we have

$$\dim |K + S + E| = a(\mathfrak{M}_n) + a(S) + a(E) + a(E \cap S) - 1. \tag{3.7}$$

The <u>arithmetic genus</u> of an arbitrary irreducible non-singular algebraic variety $\mathfrak{M}_n$ is defined as follows[14]: There exists a polynomial $v(h; \mathfrak{M}_n)$ in h such that

$$\dim |hE| = v(h; \mathfrak{M}_n) \tag{3.8}$$

for sufficiently large positive integer h, where E is a general hyperplane section of $\mathfrak{M}_n$. Then the arithmetic genus $p_a(\mathfrak{M}_n)$ of $\mathfrak{M}_n$ is defined by $p_a(\mathfrak{M}_n) = (-1)^n v(0; \mathfrak{M}_n)$. With the help of the above formulae (3.6), (3.7), we can prove[15] that the numerical characteristic $a(\mathfrak{M}_n)$ defined by (3.2) coincides with the arithmetic genus $p_a(\mathfrak{M}_n)$:

$$a(\mathfrak{M}_n) = p_a(\mathfrak{M}_n). \tag{3.9}$$

Incidentally, it follows from (3.9) that the arithmetic genus $p_a(\mathfrak{M}_n)$ is a <u>birational invariant</u> of the algebraic variety $\mathfrak{M}_n$, since each $r_\lambda(\mathfrak{M}_n)$ is known to be an birational invariant[16] of $\mathfrak{M}_n$. In view of the above identity (3.9), we define <u>the arithmetic genus of an arbitrary Kählerian variety</u> $\mathfrak{M}_n$ to be the numerical characteristic $a(\mathfrak{M}_n)$.

§4. ADJOINT SYSTEMS ON 3-DIMENSIONAL VARIETIES

In this section we consider a 3-dimensional Kählerian variety $\mathfrak{M}_3$. The system of local coordinates on $\mathfrak{M}_3$ with the center $\mathfrak{p}$ will be denoted by $z_\mathfrak{p} = (z_\mathfrak{p}^1, z_\mathfrak{p}^2, z_\mathfrak{p}^3)$. By a <u>surface</u> S on $\mathfrak{M}_3$ will be meant a compact complex analytic (reducible or irreducible) subvariety of $\mathfrak{M}_3$ of complex dimension 2. For each point $\mathfrak{p}$ on $\mathfrak{M}_3$, the surface S is represented in a neighborhood $U(\mathfrak{p})$ of $\mathfrak{p}$ by a <u>minimal local equation</u> $R_\mathfrak{p}(z_\mathfrak{p}) = 0$, where $R_\mathfrak{p}(z_\mathfrak{p})$ is a holomorphic function of $z_\mathfrak{p}$ defined

---

[13] Kodaira [5], Theorem 4.

[14] Muhley and Zariski [8], p. 82.

[15] Kodaira and Spencer [6].

[16] van der Waerden [13], §6.

in $U(\mathfrak{p})$. A singular point $\mathfrak{p}$ of S is said to be ordinary, if, by a suitable choice of the system of local coordinates $(z_{\mathfrak{p}}^1, z_{\mathfrak{p}}^2, z_{\mathfrak{p}}^3)$, the minimal local equation $R_{\mathfrak{p}} = 0$ of S at $\mathfrak{p}$ takes one of the following three forms: i) $z_{\mathfrak{p}}^1 \cdot z_{\mathfrak{p}}^2 = 0$, ii) $z_{\mathfrak{p}}^1 \cdot z_{\mathfrak{p}}^2 \cdot z^3 = 0$, iii) $(z_{\mathfrak{p}}^2)^2 - (z_{\mathfrak{p}}^1)^2 \cdot z_{\mathfrak{p}}^3 = 0$; $\mathfrak{p}$ is called a <u>double</u>, a <u>triple</u> or an (<u>ordinary</u>) <u>cuspoidal</u> point of S according as the equation $R_{\mathfrak{p}} = 0$ has the form i), ii) or iii), respectively. It is to be noted here that, in a neighborhood $U(\mathfrak{q})$ of a cuspoidal point $\mathfrak{q}$, S has the parametric representation $z_{\mathfrak{q}}^1 = u$, $z_{\mathfrak{q}}^2 = u \cdot v$, $z_{\mathfrak{q}}^3 = v^2$, where u, v are local uniformization variables on S with the center $\mathfrak{q}$. In what follows we assume that <u>the surface S has ordinary singular points only</u>.[17]

Now we construct a <u>non-singular model</u> $\tilde{S}$ of S in an obvious manner and consider S as the image $\varphi(\tilde{S})$ of the compact non-singular (possibly reducible) analytic surface $\tilde{S}$ by a holomorphic mapping φ of $\tilde{S}$ into $\mathfrak{M}_3$. It can be shown that $\tilde{S}$ has an Kählerian metric.[18] The set Δ consisting of all singular points of the surface S constitutes a (possibly reducible) compact analytic curve on $\mathfrak{M}_3$, which will be called the <u>double curve</u> of S. Clearly each triple point $\mathfrak{t}$ of S is also a triple point of Δ and Δ has no singular point other than these triple points. The inverse image $\tilde{\Delta} = \varphi^{-1}(\Delta)$ is a compact analytic curve on $\tilde{S}$. Incidentally we denote by $\{\mathfrak{t}\}$ (or $\{\mathfrak{q}\}$) the set consisting of all triple points $\mathfrak{t}$ (or of all cuspoidal points $\mathfrak{q}$) of S. Take a triple point $\mathfrak{t} \in \{\mathfrak{t}\}$ and denote by $S^{(1)}$, $S^{(2)}$, $S^{(3)}$ three sheets of S passing through $\mathfrak{t}$. The inverse image $\varphi^{-1}(\mathfrak{t})$ consists of three points τ_1, τ_2, τ_3 on S corresponding respectively to $S^{(1)}, S^{(2)}, S^{(3)}$. On the other hand, there exist three branches $\mathfrak{t}_1 = S^{(2)} \cap S^{(3)}$, $\mathfrak{t}_2 = S^{(3)} \cap S^{(1)}$, $\mathfrak{t}_3 = S^{(1)} \cap S^{(2)}$ of Δ passing through $\mathfrak{t}$ and, for each $\mathfrak{t}_\lambda$, the inverse image $\varphi^{-1}(\mathfrak{t}_\lambda)$ consists of two branches $\tilde{\mathfrak{t}}_\lambda'$, $\tilde{\mathfrak{t}}_\lambda''$ of $\tilde{\Delta}$. We arrange these six branches $\tilde{\mathfrak{t}}_\lambda^{(m)}$ $(\lambda = 1,2,3,\ m = 1,2)$ in such a way that $\tilde{\mathfrak{t}}_\lambda^{(m)}$ passes through τ_k if $\lambda + m \equiv k(3)$. Clearly each $\tau_k \in \varphi^{-1}(\mathfrak{t})$, $\mathfrak{t} \in \{\mathfrak{t}\}$ is a double point of $\tilde{\Delta}$ and $\tilde{\Delta}$ has no singular points other than these double points. A <u>place</u> on a compact analytic curve is, by definition, a pair $[p, \mathfrak{b}]$ of a point p on the curve and a branch $\mathfrak{b}$ of the curve passing through p. We denote the places $[\mathfrak{t}, \mathfrak{t}_\lambda]$, $[\tau_k, \tilde{\mathfrak{t}}_\lambda^{(m)}]$ simply by $\mathfrak{t}_\lambda$, $\tilde{\mathfrak{t}}_\lambda^{(m)}$, respectively. The set of all places on Δ (or $\tilde{\Delta}$) constitutes the <u>non-singular model</u> of Δ (or $\tilde{\Delta}$) which will be denoted by the same symbol Δ (or $\tilde{\Delta}$). Clearly φ can be regarded as a holomorphic mapping of the non-singular model $\tilde{\Delta}$ onto the non-singular model Δ. The inverse image $\varphi^{-1}(\mathfrak{p})$ of each place $\mathfrak{p}$ of

---

[17] We assume this in view of the following well known fact: Every algebraic surface can be birationally transformed into a surface in a 3-dimensional projective space with ordinary singularities only.

[18] Kodaira [4], §1.

Δ consists of two places $\tilde{\mathfrak{p}}'$, $\tilde{\mathfrak{p}}''$ of $\tilde{\Delta}$ unless $\mathfrak{p} \in \{\mathfrak{q}\}$, while, for each $\mathfrak{q} \in \{\mathfrak{q}\}$, $\varphi^{-1}(\mathfrak{q})$ consists of only one place $\tilde{\mathfrak{q}} = \varphi^{-1}(\mathfrak{q})$. Thus (the non-singular model of) $\tilde{\Delta}$ is a two-fold covering manifold of (the non-singular model of) Δ having each $\tilde{\mathfrak{q}} = \varphi^{-1}(\mathfrak{q})$, $\mathfrak{q} \in \{\mathfrak{q}\}$, as a simple branch point. We denote by J the covering transformation of $\tilde{\Delta}$ with respect to Δ which interchanges two sheets of $\tilde{\Delta}$. Let

$$S = \sum_{\lambda=1}^{\mu} S_\lambda$$

or

$$\Delta = \sum_{j=1}^{s} \Delta_j$$

be the decomposition of S or Δ into its irreducible components S_λ or Δ_j, respectively. Each irreducible component Δ_j is said to be of the first or of the second kind according as its inverse image $\tilde{\Delta}_j = \varphi^{-1}(\Delta_j)$ is reducible or irreducible. In case Δ_j is of the first kind, $\tilde{\Delta}_j$ consists of two irreducible components $\tilde{\Delta}_j'$, $\tilde{\Delta}_j'' = J\tilde{\Delta}_j'$, since $\tilde{\Delta}_j$ is a two-fold covering space of Δ_j. For each irreducible component Δ_j of the first kind, we define the <u>index</u> $(\mathfrak{t} \cdot \Delta_j)$ of Δ_j at each triple point $\mathfrak{t}$ by

$$(\mathfrak{t} \cdot \Delta_j) = \sum_{\lambda, m} (-1)^m \,\delta(\tilde{\mathfrak{t}}_\lambda^{(m)}, \tilde{\Delta}_j') \tag{4.1}$$

where $\delta(\tilde{\mathfrak{t}}_\lambda^{(m)}, \tilde{\Delta}_j')$ is 1 or 0 according as $\tilde{\mathfrak{t}}_\lambda^{(m)} \in \tilde{\Delta}_j'$ or $\notin \tilde{\Delta}_j'$.

Now we shall introduce several numerical characteristics of S. First, we define the <u>virtual arithmetic genus</u> $a(S)$ of S by

$$a(S) = t + \tfrac{1}{2} c + \sum_j (\pi_j - 1) + \sum_\lambda (a_\lambda + 1) - 1, \tag{4.2}$$

where t is the number of triple points of S, c is the number of cuspoidal points of S, π_j is the genus of Δ_j, $a_\lambda = a(\tilde{S}_\lambda)$ is the arithmetic genus of the non-singular model $\tilde{S}_\lambda = \varphi^{-1}(S_\lambda)$, and the sum $\sum_j$ or $\sum_\lambda$ is to be extended over all irreducible components Δ_j of Δ or S_λ of S, respectively. Second, for an arbitrary simple differential A [or α] of the first kind on $\mathfrak{W}_3$ [or $\tilde{S}$], we denote by $A_{\tilde{S}}$ [or $\alpha_{\tilde{\Delta}}$] the simple differential on $\tilde{S}$ [or $\tilde{\Delta}$] induced by A [or α]. It is obvious that, if $\alpha = A_{\tilde{S}}$, then $\alpha_{\tilde{\Delta}}$ satisfies $J\alpha_{\tilde{\Delta}} = \alpha_{\tilde{\Delta}}$; thus the linear space $A_{\tilde{S}}$ consisting of all induced differentials $A_{\tilde{S}}$ is a subspace of the linear space $\{\alpha \mid J\alpha_{\tilde{\Delta}} = \alpha_{\tilde{\Delta}}\}$. Now we define the <u>deficiency</u> δ of S <u>with respect to simple differentials</u> by

$$\delta = \dim\{\alpha \mid J\alpha_{\tilde{\Delta}} = \alpha_{\tilde{\Delta}}\} - \dim\{A_{\tilde{S}}\}. \tag{4.3}$$

Third, we associate with each irreducible component Δ_j of the first kind an indeterminate x_j and consider the simultaneous linear equations

$$(4.4) \qquad \sum_j (\mathfrak{t} \cdot \Delta_j) x_j = 0, \quad \text{for all triple points } \mathfrak{t},$$

where the sum $\sum_j$ is to be extended over all irreducible components Δ_j of the first kind. Then it can be shown[19] that the number η^* of linearly independent solutions x_j of (4.4) is not smaller than the difference $\mu - m$ between the number μ of irreducible components of S and the number m of connected components of S; thus the integer

$$(4.5) \qquad \eta = \eta^* - \mu + m$$

is non-negative. We call this integer η <u>the deficiency of the singularities</u> of S.

THEOREM 5.[20] The dimension of the adjoint system $|K + S|$ of S is given by

$$(4.6) \qquad \dim |K + S| = a(\mathfrak{M}_3) + a(S) + \ell - k + \delta + \eta - m,$$

where $A(\mathfrak{M}_3)$ is the arithmetic genus of $\mathfrak{M}_3$, $a(S)$ is the virtual arithmetic genus of S, δ is the deficiency of S with respect to simple differentials, η is the deficiency of the singularities of S, m is the number of connected components of S and ℓ or k is respectively the number of linearly independent double differentials B or simple differentials A of the first kind on $\mathfrak{M}_3$ which vanish on S in the sense that $B_{\tilde{S}}$ or $A_{\tilde{S}}$ vanish identically, $B_{\tilde{S}}$ or $A_{\tilde{S}}$ being the differential on the non-singular model $\tilde{S}$ of S induced by B or A, respectively.

We shall give here an outline[21] of a proof of the above Theorem 5. Let $\mathfrak{w}(\tilde{\Delta})$ be the linear space consisting of all double differentials w on $\tilde{S}$ which are multiples of $-\tilde{\Delta}$. We associate with each $w \in \mathfrak{w}(\tilde{\Delta})$ the 4-current $T(w)$ of type $(3,1)$ on $\mathfrak{M}_3$ defined by $T(w)[\Phi] = \int_{\tilde{S}} w \cdot \Phi_{\tilde{S}}$, where $\Phi_{\tilde{S}}$ is the 2-form induced by Φ on $\tilde{S}$. Each $W \in \mathfrak{W}(S)$ can be considered as a 3-current of type $(3,0)$ on $\mathfrak{M}_3$, since

[19] Kodaira [4], §6.

[20] Kodaira [4], Theorem III.

[21] For a detailed proof, see Kodaira [4], §1-7.

the integral
$$W[\psi] = \int_{\mathfrak{M}_3} W \cdot \psi$$

converges absolutely for an arbitrary 3-form ψ of class C^∞. It can be shown that the exterior derivative of this current W is represented in the form $dW = 2\pi i T(w)$, where $w \in \mathfrak{w}(\tilde{\Delta})$. In the same manner as in the proof of Theorem 2 above, we can show that, for given $w \in \mathfrak{w}(\tilde{\Delta})$, there exists $W \in \mathfrak{W}(S)$ with $dW = 2\pi i T(w)$ if and only if $T(w)$ satisfies $dT(w) = HT(w) = 0$. On the other hand dW is equal to zero if and only if W is of the first kind. Setting

$$\mathfrak{w}_0 = \{w \mid w \in \mathfrak{w}(\tilde{\Delta}), \quad dT(w) = HT(w) = 0\}, \tag{4.7}$$

we infer therefore that

$$\dim \mathfrak{W}(S) = r_3(\mathfrak{M}_3) + \dim \mathfrak{w}_0. \tag{4.8}$$

The divisor $\mathfrak{c} = \sum_{\lambda,m} \tilde{\mathfrak{t}}_\lambda^{(m)}$ on $\tilde{\Delta}$ is called the conductor of the curve $\tilde{\Delta}$ on $\tilde{S}$. Denote by $\Xi(\mathfrak{c})$ the linear space consisting of all differentials ξ on $\tilde{\Delta}$ which are multiples of $-\mathfrak{c}$ and associate with each $\xi \in \Xi(\mathfrak{c})$ the 3-current $T(\xi)$ of type $(2,1)$ on $\tilde{S}$ defined by $T(\xi)[\psi] = \int_{\tilde{\Delta}} \xi \cdot \psi_{\tilde{\Delta}}$, where ψ is a "variable" 1-form of class C^∞ on $\tilde{S}$ and $\psi_{\tilde{\Delta}}$ is the 1-form induced by ψ on $\tilde{\Delta}$. Then each double differential $w \in \mathfrak{w}(\tilde{\Delta})$ determines its <u>residue</u> $\xi \in \Xi(\mathfrak{c})$ which is related with w by the relation $dw = 2\pi i T(\xi)$. We denote the residue ξ of w by $\mathfrak{R}(w)$. It can be shown[21a] that, for given $\xi \in \Xi(\mathfrak{c})$, there exists $w \in \mathfrak{w}(\tilde{\Delta})$ with $dw = 2\pi i T(\xi)$ if and only if ξ satisfies

$$\int_{\tilde{\Delta}} \xi \cdot \alpha_{\tilde{\Delta}} = 0 \tag{4.9}$$

for all simple differentials α of the first kind on $\tilde{S}$ and

$$\mathrm{Res}\,[\tilde{\mathfrak{t}}''_\lambda](\xi) + \mathrm{Res}\,[\tilde{\mathfrak{t}}'_\nu](\xi) = 0 \tag{4.10}$$

for each triple point $\mathfrak{t}$ of S, where $(\lambda, \nu) = (2,3), (3,1)$ or $(1,2)$ and $\mathrm{Res}\,[\tilde{\mathfrak{t}}_\lambda^{(m)}](\xi)$ denotes the residue of ξ at the place $\tilde{\mathfrak{t}}_\lambda^{(m)}$. It is to be noted here that $\tilde{\mathfrak{t}}''_\lambda$, $\tilde{\mathfrak{t}}'_\nu$ appearing in (4.10) corresponds to one and the same double point of $\tilde{\Delta}$. Thus the linear mapping $w \longrightarrow \xi = \mathfrak{R}(w)$ maps $\mathfrak{w}(\tilde{\Delta})$ onto the subspace Ξ of $\Xi(\mathfrak{c})$ consisting of all $\xi \in \Xi(\mathfrak{c})$ satisfying (4.9) and (4.10). Clearly the kernel of the mapping $w \longrightarrow \xi = \mathfrak{R}(w)$ is the space consisting of all double differentials of the first kind on $\tilde{S}$ which will be denoted by $\mathfrak{w}(0)$. Now it is easy to see that the first condition $dT(w) = 0$ is equivalent to $J\,\mathfrak{R}(w) + \mathfrak{R}(w) = 0$, while the second condition $HT(w) = 0$ is equivalent to

[21a] Kodaira [3], pp. 860-863.

(4.11)
$$\int_{\widetilde{S}} \overline{w} \cdot B_{\widetilde{S}} = 0$$

for all double differentials B of the first kind on $\mathfrak{M}_3$. Moreover, since each $B_{\widetilde{S}}$ is contained in $\mathfrak{w}(0)$, the condition (4.11) is independent of $J\mathfrak{R}(w) + \mathfrak{R}(w) = 0$. Hence, denoting by $\Xi^{(-)}$ the subspace of Ξ consisting of all $\xi \in \Xi$ satisfying

(4.12)
$$J\xi + \xi = 0,$$

we infer that the mapping $w \rightarrow \xi = \mathfrak{R}(w)$ maps $\mathfrak{w}_0$ onto $\Xi^{(-)}$ and that the kernel $\mathfrak{w}_0 \cap \mathfrak{w}(0)$ of this mapping consists of all double differentials w of the first kind on $\widetilde{S}$ satisfying (4.11). Clearly (4.11) involves $r_2(\mathfrak{M}_3) - \ell$ linearly independent conditions. Consequently, setting $g = \dim \mathfrak{w}(0) = \sum_\lambda r_2(\widetilde{S}_\lambda)$, we get $\dim [\mathfrak{w}_0 \cap \mathfrak{w}(0)] = g - r_2(\mathfrak{M}_3) + \ell$ and therefore

(4.13)
$$\dim \mathfrak{w}_0 = g - r_2(\mathfrak{M}_3) + \ell + \dim \Xi^{(-)}.$$

Now it is not difficult to count the dimension of the linear space $\Xi^{(-)}$. Let $|\xi|^{(-)}$ be the subspace of $\Xi(\mathfrak{c})$ consisting of all differentials $\xi \in \Xi(\mathfrak{c})$ satisfying (4.10) and (4.12); again let $|\sigma|^{(-)}$ be the space consisting of all differentials σ of the first kind on $\widetilde{\Delta}$ satisfying $J\sigma + \sigma = 0$. Assume that $\Delta_1, \Delta_2, \ldots, \Delta_f$ are irreducible components of the first kind and that $\Delta_{f+1}, \ldots, \Delta_s$ are of the second kind. Obviously each cuspoidal point $\mathfrak{q}$ of S lies on one of Δ_j, $j > f$. Denote c_j the number of cuspoidal points $\mathfrak{q}$ lying on Δ_j. Then $\widetilde{\Delta}_j = \varphi^{-1}(\Delta_j)$, $j > f$, is a two-fold covering space of Δ_j having c_j simple branch points. Using this fact, we infer readily that

(4.14)
$$\dim |\sigma|^{(-)} = \sum_{j=1}^{s} (\pi_j - 1) + f + \frac{1}{2} \sum_{j=f+1}^{s} c_j.$$

For each $\xi \in |\xi|^{(-)}$, the quantity $\rho(\mathfrak{t}) = (-1)^m \operatorname{Res}[\widetilde{\mathfrak{t}}_\lambda^{(m)}](\xi)$ depends only on $\mathfrak{t}$. Moreover we have

(4.15)
$$\sum_{\mathfrak{t}} (\mathfrak{t} \cdot \Delta_j) \rho(\mathfrak{t}) = 0, \quad (j = 1, 2, \ldots, f).$$

Conversely, for given $\rho(\mathfrak{t})$ satisfying (4.15), there exists $\xi \in |\xi|^{(-)}$ with $\operatorname{Res}[\mathfrak{t}_\lambda^{(m)}](\xi) = (-1)^m \rho(\mathfrak{t})$. This ξ is determined uniquely modulo $|\sigma|^{(-)}$. Denoting by e the rank of the simultaneous equations (4.15), we infer therefore that

(4.16)
$$\dim |\xi|^{(-)} = t - e + \dim |\sigma|^{(-)}.$$

Now, since $J\xi = -\xi$ for any $\xi \in |\xi|^{(-)}$, the condition (4.9) is equivalent to the following one:

$$\int_{\Delta} \overline{\xi} \cdot (\alpha_{\Delta} - J\alpha_{\Delta}) = 0 \tag{4.17}$$

for all simple differentials α of the first kind on $\tilde{S}$. Clearly (4.17) involves $q - \delta - r_1(\mathfrak{M}_3) + k$ independent conditions, where $q = \sum_{\lambda} r_1(\tilde{S}_{\lambda})$. Hence we get

$$\dim \Xi^{(-)} = \dim |\xi|^{(-)} - q + \delta + r_1(\mathfrak{M}_3) - k. \tag{4.18}$$

Since (4.15) is the transposed system of (4.4), we have $\eta^* = f - e$ and therefore $e = f - \eta - \mu + m$. Inserting this into (4.16) and combining it with (4.14) and (4.18), we get

$$\dim \Xi^{(-)} = t + \frac{1}{2} c + \sum_{j=1}^{s} (\pi_j - 1) + \mu - q + r_1(\mathfrak{M}_3) - k + \delta + \eta - m.$$

Now, inserting this into (4.13) and combining it with (4.8), we obtain the formula (4.6), q. e. d.

In what follows we assume $\mathfrak{M}_3$ to be a 3-dimensional algebraic variety without singularities imbedded in a projective space. Denote by E a <u>general</u> hyperplane section of $\mathfrak{M}_3$. Then it is easy to show[22] that $\ell = k = \delta = \eta = 0$ and $m = 1$ for any surface of the type $S = S' + E$, S' being a surface with ordinary singularities only. Thus we obtain from Theorem 5 the following

THEOREM 6. In case the surface S is of the form $S = S' + E$, E being a general hyperplane section of $\mathfrak{M}_3$ (the case in which $S = E$ is included), we have

$$\dim |K + S| = a(\mathfrak{M}_3) + a(S) - 1. \tag{4.19}$$

The 4-cycle $-K$ represents the 3rd basic characteristic class[23] of $\mathfrak{M}_3$. We denote by C the 2-cycle expressing the 2nd basic characteristic class of $\mathfrak{M}_3$. We remark here that we can choose as C an algebraic cycle. Now, the virtual airthmetic genus $a(S)$ of S defined by (4.2) can be represented in the form

$$a(S) = \frac{1}{6} S^3 + \frac{1}{4} KS^2 + \frac{1}{12} (K^2 + C)S - 1, \tag{4.20}$$

---

[22] Kodaira [4], §8.

[23] Chern [1], pp. 99-103. From the point of view of the theory of currents, each homology class can be identified with its dual cohomology class. See de Rham and Kodaira [9], pp. 22-42.

where S^3, KS^2, ... denote the topological intersections numbers $I(S,S,S)$, $I(K,S,S)$, ..., respectively. This formula (4.20) can be proved as follows: First, consider the case in which S is an irreducible non-singular surface. Denote by χ_S the Euler characteristic of S and by K_S the canonical divisor on S. Then we have

$$(4.21)^{24} \qquad 12(a(S) + 1) = \chi_S + K_S^2 ,$$

where K_S^2 denotes the intersection number $I_S(K_S, K_S)$ of the 2-cycle K_S on S with itself. Using the so-called duality theorem on complex sphere bundles, we can prove the adjunction formulae[25]

$$K_S \backsim KS + S^2, \quad \text{on } S,$$
$$\chi_S = CS + I_S(K_S, S^2),$$

where S^2 denotes the divisor $D \cdot S$ on S cute out by a "general" divisor $D \approx S$. Inserting these formulae into (4.21), we get immediately (4.20). Now, the general case in which S is a surface with ordinary singularities can be readily reduced to the above special case by using the fact that the complete linear system $|S + hE|$ contains an irreducible non-singular surface if h is sufficiently large.

In view of (4.20), we introduce the following definition: By the virtual airthmetic genus of an arbitrary divisor D on $\mathfrak{M}_3$ we shall mean the integer $a(D)$ defined by

$$(4.22) \qquad a(D) = \frac{1}{6} D^3 + \frac{1}{4} KD^2 + \frac{1}{12} (K^2 + C)D-1.$$

It is convenient to introduce here the binary functional

$$\pi(D_1D_2) = \frac{1}{2} (D_1^2 D_2 + D_1 D_2^2 + KD_1 D_2) + 1.$$

Then we have

$$(4.23) \qquad a(D_1 + D_2) = a(D_1) + a(D_2) + \pi(D_1D_2).$$

In case S is an irreducible non-singular surface on $\mathfrak{M}_3$, $\pi(DS)$ represents the virtual genus of the divisor DS on S cut out by D, provided that D does not contain S as one of its components. Following Severi, we call $\pi(D^2) = \pi(DD)$ the virtual curve genus of D.

Now, consider the adjoint system $|D| = |K + S|$ of an arbitrary surface S with ordinary singularities only. Then we get easily

$$a(S) = D^3 - \pi(D^2) + a(D) + 1 - \frac{1}{12} KC.$$

[24] This is a consequence of a formula of Noether. See Zariski [16], p. 62, p. 113.

[25] Hodge [2]. For a proof of these formulae due to Chern based on the duality theorem on complex sphere bundles, see Kodaira [4], §9.

Hence, setting

$$p_a(\mathfrak{M}_3) = \frac{1}{12} KC - a(\mathfrak{M}_3) + 2, \tag{4.24}$$

we obtain from (4.6) the following

THEOREM 7. The dimension of the adjoint system $|D| = |K + S|$ of an arbitrary surface S on $\mathfrak{M}_3$ with only ordinary singularities is given by

$$\begin{aligned} \dim |D| = D^3 - \pi(D^2) + a(D) - \\ - p_a(\mathfrak{M}_3) + 3 - m + \ell - k + \delta + \eta, \end{aligned} \tag{4.25}$$

where m, ℓ, k, δ, η have the same meaning as in Theorem 5.

It is easy to see that, if S is irreducible and if $\dim |S| \geqq 1$, the constant k appearing in (4.25) must be equal to zero. Hence we get from (4.25) the following theorem due to Severi[26]:

THEOREM 8. If S is an irreducible surface with ordinary singularities only and if $\dim |S| \geqq 1$, then the dimension of the adjoint system $|D| = |K + S|$ satisfies the inequality

$$\dim |D| \geqq D^3 - \pi(D^2) + a(D) - p_a(\mathfrak{M}_3) + 2. \tag{4.26}$$

Again it follows from Theorem 6 the following

THEOREM 9. In case a complete linear system $|D|$ on $\mathfrak{M}_3$ is sufficiently ample in the sense that $|D - K - E|$ contains a surface with ordinary singularities only, E being a general hyperplane section of $\mathfrak{M}_3$ (the case in which $|D| = |K + E|$ is included), the dimension of $|D|$ is given by

$$\dim |D| = D^3 - \pi(D^2) + a(D) - p_a(\mathfrak{M}_3) + 2. \tag{4.27}$$

The above formula (4.27) is also written in the form

$$\dim |D| = \frac{1}{6} D^3 - \frac{1}{4} KD^2 + \frac{1}{12}(K^2 + C)D - p_a(\mathfrak{M}_3).$$

This shows that $p_a(\mathfrak{M}_3)$ is the arithmetic genus of $\mathfrak{M}_3$ so that $p_a(\mathfrak{M}_3) = a(\mathfrak{M}_3)$. Consequently we get from (4.24) the following relation[27]:

---

[26] Severi [10], p. 63.

[27] Todd [12], p. 215.

$$(4.28) \qquad a(\mathfrak{M}_3) = p_a(\mathfrak{M}_3) = \frac{1}{24} KC + 1.$$

The index of speciality $i\,|D|$ of an arbitrary complete linear system $|D|$ is defined by $i\,|D| = \dim |K - D| + 1$. Now, setting

$$\dim |D| = D^3 - \pi(D^2) + a(D) - p_a(\mathfrak{M}_3) + 2 + i\,|D| + \sup |D| ,$$

we introduce the <u>superabundance</u> $\sup |D|$ of $|D|$. Then it follows from (4.25) that the superabundance of the adjoint system $|K + S|$ of an arbitrary surface S with only ordinary singularities is given by

$$(4.29) \qquad \sup |K + S| = \ell - k + \delta + \eta + 1 - m.$$

The above Theorem 9 shows that the superabundance of a "sufficiently ample" system $|D|$ is always equal to zero.

§5. EXAMPLES

In this section we assume $\mathfrak{M}_3$ to be the direct product of an algebraic curve $\mathfrak{C}$ and an algebraic surface $\mathfrak{Z}$: $\mathfrak{M}_3 = \mathfrak{C} \times \mathfrak{Z}$ and consider two examples of surfaces S on $\mathfrak{M}_3$ whose adjoint systems are not "sufficiently ample". Setting $\pi = r_1(\mathfrak{C})$, $g = r_2(\mathfrak{Z})$, $q = r_1(\mathfrak{Z})$, $a = a(\mathfrak{Z}) = r_2(\mathfrak{Z}) - r_1(\mathfrak{Z})$, we get readily $r_1(\mathfrak{M}_3) = \pi + q$, $r_2(\mathfrak{M}_3) = \pi \cdot q + g$, $r_3(\mathfrak{M}_3) = \pi \cdot g$ and therefore

$$a(\mathfrak{M}_3) - 1 = (\pi - 1)(a + 1).$$

Again, denoting by $\mathfrak{k}$ or $K_{\mathfrak{Z}}$ the canonical divisor on $\mathfrak{C}$ or $\mathfrak{Z}$, respectively, we infer readily that the canonical divisor K on $\mathfrak{M}_3$ is given by $K = \mathfrak{k} \times \mathfrak{Z} + \mathfrak{C} \times K_{\mathfrak{Z}}$.

1) Take a divisor $\mathfrak{d} = \sum_{\nu=1}^{m} \mathfrak{p}_\nu$ on $\mathfrak{C}$ composed of m different points $\mathfrak{p}_1, \mathfrak{p}_2, \ldots, \mathfrak{p}_m$ and set $S = \mathfrak{d} \times \mathfrak{Z}$. Then we get $a(S) = ma + m - 1$ and therefore

$$a(\mathfrak{M}_3) + a(S) = (a + 1)(m + \pi - 1).$$

Furthermore we infer readily that $\ell = \pi q$, $k = \pi$, $\delta = (m - 1)q$, and $\eta = 0$. Using (4.29), we get therefore

$$(5.1) \qquad \sup |K + S| = (q - 1)(m + \pi - 1).$$

Consequently we obtain

$$(5.2) \qquad \dim |K + S| + 1 = g(m + \pi - 1).$$

It is to be noted here that the second factor $m + \pi - 1$ of the right hand side of (5.2) is equal to the dimension of the complete linear series $|\mathfrak{k} + \mathfrak{d}|$ on $\mathfrak{C}$ increased by 1. In case $q(= r_1(\mathfrak{Z})) = 0$, it follows from (5.1) that $\sup |K + S| = - m - \pi + 1$. This shows that the superabundance $\sup |K + S|$ is not necessarily non-negative.

ii) Take an irreducible curve Γ on $\mathfrak{Z}$ with ordinary double points only and set $S = \mathfrak{C} \times \Gamma$. Then, denoting by p the genus of Γ and by d the number of double points of Γ, we get $a(S) = (\pi-1)(p+d-1) - 1$ and therefore

$$a(\mathfrak{M}_3) + a(S) = (\pi - 1)(p + d + a).$$

Denote by k_Γ the number of linearly independent simple differentials of the first kind on $\mathfrak{Z}$ vanishing on Γ. Then we get $\ell = \pi k_\Gamma + g$, $k = k_\Gamma$, $\delta = p - q + k_\Gamma$, $\eta = d$, $m = 1$, and therefore

$$\sup |K + S| = p + d + a + \pi k_\Gamma .$$

Consequently we obtain

$$\dim |K + S| + 1 = \pi(p + d + a + k_\Gamma). \tag{5.3}$$

We remark that the second factor $p + d + a + k_\Gamma$ of the right hand side of (5.3) equals the dimension of the complete linear system $|K_{\mathfrak{Z}} + \Gamma|$ on $\mathfrak{Z}$ increased by 1.

BIBILIOGRAPHY

[1] CHERN, S., "Characteristic classes of Hermitian manifolds," Annals of Math., vol. 47 (1946), pp. 85-121.

[2] HODGE, W. V. D., "The characteristic classes on algebraic varieties," Proc. London Math. Soc., Ser. 3, vol. I (1951), pp. 138-151.

[3] KODAIRA, K., "The theorem of Riemann-Roch on compact analytic surfaces," Amer. Jour. of Math. vol. 73 (1951), pp. 813-875.

[4] ——————,"The Theorem of Riemann-Roch for adjoint systems on 3-dimensional algebraic varieties," Annals of Math., vol. 56 (1952), pp. 298-342.

[5] ——————, "On the theorem of Riemann-Roch for adjoint systems on Kählerian varieties," Proc. Nat. Acad. Sciences, vol. 38 (1952), pp. 522-527.

[6] KODAIRA, K., and SPENCER, D. C., "On Arithmetic genera of algebraic varieties," to appear in Proc. Nat. Acad. Sciences.

[7] LEFSCHETZ, S., "L'Analysis situs et la geometrie algebrique," Paris (1924).

[8] MUHLEY, H T., ZARISKI, O., "Hilbert's characteristic function and the arithmetic genus of an algebraic variety," Trans. Amer. Math. Soc., vol. 69 (1950), pp. 78-88.

[9] de RHAM, G., KODAIRA, K., "Harmonic integrals," (Mimeographed notes), Institute for Advanced Study, Princeton (1950).

[10] SEVERI, F., "Fondamenti per la geometria sulle varieta algebriche," Rend. Circ. Math. Palermo, vol. 28 (1909), pp. 33-87.

[11] SPENCER, D. C., "Green's operators on manifolds," these Proceedings.

[12] TODD, J. A., "The arithmetic invariants of algebraic loci," Proc. London Math. Soc. 2nd Series, vol. 43 (1937), pp. 190-225.

[13] van der WAERDEN, B. L., "Birational invariants of algebraic manifolds," Acta Salmanticensia, vol. II (1947).

[14] WEIL, A., "Sur la theorie des formes differentielles attachees a une variete analytique complexe," Comm. Math. Helv., vol. 20 (1947), pp. 110-116.

[15] WEYL, H., "Die Idee der Riemannschen Fläche," Berlin (1913).

[16] ZARISKI, O., "Algebraic surfaces," Ergeb. Math. Grenz., vol. 3, no. 5 (1935).

Lightning Source UK Ltd.
Milton Keynes UK
UKHW040618200522
403285UK00001B/52